BLOWING BUBBLES IN THE COSMOS

BLOWING BUBBLES IN THE COSMOS

Astronomical Winds, Jets, and Explosions

T. W. Hartquist
J. E. Dyson
D. P. Ruffle

UNIVERSITY PRESS

2004

Oxford   New York
Auckland   Bangkok   Buenos Aires   Cape Town   Chennai
Dar es Salaam   Delhi   Hong Kong   Istanbul   Karachi   Kolkata
Kuala Lumpur   Madrid   Melbourne   Mexico City   Mumbai   Nairobi
São Paulo   Shanghai   Taipei   Tokyo   Toronto

Published by Oxford University Press, Inc.
198 Madison Avenue, New York, New York, 10016

www.oup.com

Oxford is a registered trademark of Oxford University Press

Library of Congress Cataloging-in-Publication Data
Hartquist, T. W.
Blowing bubbles in the cosmos : astronomical winds, jets, and explosions / T. W.
Hartquist, J. E. Dyson, and D. P. Ruffle.
p.   cm.
Includes bibliographical references and index.
ISBN 0-19-513054-5
1. Solar wind.   2. Astrophysical jets.   3. Stars, New.   I. Dyson, J. E. (John Edward),
1941–   II. Ruffle, D. P.   III. Title.
QB529.H37 2003
523.01—dc21   2003051706

9   8   7   6   6   5   4   3   2   1

Printed in the United States of America
on acid-free paper

We are all in the gutter, but some of us are looking at the stars.
—Oscar Wilde

PREFACE

Quite commonly, a book on astronomy concerns a single object or class of objects. For instance, stars constitute a perfectly reasonable subject for a book, and volumes on the Big Bang or some aspect of it proliferate. In practice, a typical astrophysicist does not restrict research to one type of astronomical object or to one cosmological epoch. Rather, he or she adapts insight obtained from the study of one variety of environment to make progress in the investigation of another. An astrophysicist's flexibility and success in attaining knowledge of the physical world are possible only because the structures and dynamics of widely different entities are controlled by the same handful of physical processes and are describable with a fairly small number of simple pictures.

We have chosen not to concentrate on a single class of radiation sources, such as young stars, evolved stars, galaxies, or active galactic nuclei. Instead, we have elected to cover a great number of astrophysical environments, from the cosmological intergalactic medium to regions around supermassive black holes as well as clouds in which the youngest stars have just formed. We do so by focussing on winds, bubbles, jets, and explosions. The underlying pictures required for an examination of such phenomena are simple, limited in type, and highly adaptable. Thus, we introduce a large fraction of astrophysics from a unifying perspective.

We suspect that many readers of this volume will have studied high school physics. The book should provide easy reading for a professional scientist or engineer. However, we hope that it will be accessible to motivated readers who have no advanced training in science or technology but who are curious about the mechanisms at work in the Universe. For instance, we would be very pleased if young people preparing in school to undertake university study in science would find our book interesting. To assist such readers we have included treatments of some physics and occasional asides on notation and terms that others may wish to skip. We have also provided a glossary and avoided equations in the main text. For those who are more mathematically inclined, we have included an appendix containing some of the key formulae.

Winds, explosions, and the bubbles that they blow are ubiquitous in the cosmos. Some of them are associated with the most energetic phenomena occurring at present in the Universe. Some produce a large fraction of the most spectacular images made with the Hubble Space Telescope or indeed any telescope.

The nearest example of an astronomical wind is the solar wind. Its impact on the Earth's upper atmosphere induces aurorae. Chapter 1 begins with a brief history of early auroral research. More generally the chapter describes how astronomical winds were first discovered. Somewhat surprisingly, the existence of the continuous winds of some other stars was established before the presence of such an outflow from the Sun was demonstrated.

Chapter 2 provides a frame of reference for the reader when considering the properties of astronomical sources throughout the volume. It covers briefly the masses, timescales, densities, and temperatures associated with astronomical phenomena and objects.

The first astronomical winds to be discovered were those associated with individual stars. Many other windy sources contain stars. Chapter 3 is a continuation of the presentation of background material important for the understanding of the remainder of the book. It contains an introduction to the evolution of stars. The success of scientists in accounting for the observed relationship between a star's temperature and its brightness is one of the most remarkable intellectual accomplishments of the 20th century. The calculations show that the evolution of a star in its late phases depends critically on the rate at which the stellar wind transports mass away.

Hydrodynamics is the science of the bulk motions or flows of nonmagnetized gases and fluids and is central to much of modern day astrophysics. Chapter 4 is an introductory exposition of hydrodynamics and its application to the structure and dynamics of a wind from an astronomical source. The basic structure of a bubble blown by a supersonic wind affecting a medium surrounding the source is treated in this chapter.

Young stellar objects with masses comparable to that of the Sun constitute the subject of chapter 5. Rotation and magnetic fields play significant roles in the creation of winds and jets in some of these sources. The rotation is a consequence of the processes leading to the formation of the sources. Thus, the birth of the young stars is described at some length so that the reader will know the origin of the conditions giving rise to the mass loss and the confinement of flow to narrowed ranges of directions. We refer to such confinement as "collimation."

Chapter 6 concerns mass loss from and bubbles blown by young stellar objects that each contain at least eight times more mass than the Sun. The high radiative luminosities of such objects drive bubbles that differ qualitatively from wind-blown bubbles. Masers, long-wavelength versions of lasers, are found near the bubbles around many such objects. Emission seen from molecules moving at speeds of about 100 kilometers per second exists as a result of the interactions of the winds from the massive young stars with the surrounding media.

Chapter 7 contains descriptions of mechanisms responsible for driving mass loss from stars that have evolved beyond the stages addressed in chapters 5 and 6. Processes examined include the action of thermal pressure, radiation pressure on gas or dust, and the propagation of waves in magnetized material.

A star containing more than about eight times the mass of the Sun will evolve to cause a spectacular explosion called a supernova. Chapter 8 concerns supernova explosions and supernova remnants, which are the bubbles blown by these explosions. We describe the use of supernovae to study the possibilities that the expansion of the Universe is accelerating and that the general relativistic equations governing gravity contain a "cosmological constant." Cosmic rays are particles moving at a good fraction of the speed of light. We address their acceleration in supernova remnants.

In some galaxies, impressively powerful bursts of star formation occur, causing concentrations of supernovae and the production of galactic superwinds. Chapter 9 is about these starbursts and superwinds. It also touches on the ways in which winds may have affected structure formation in the Early Universe as well as on the acceleration of cosmic rays in galactic winds.

The bursts of star formation mentioned above take place within the central few hundred light years of the galaxies in which they occur. Even more compact and more powerful sources known as active galactic nuclei are at the focus of chapter 10. Gas moving at thousands of kilometers per second and more exists at fractions of a light year from the supermassive black holes that power these objects.

Chapter 11 contains short descriptions of a number of types of windy and explosive sources not treated in detail in other chapters. These sources include the gamma-ray bursts which are now known to be extragalactic in origin.

An epilogue concludes the volume.

At the ends of chapters we have included lists of selected references. Many of these are of a more advanced nature than some readers will find suitable. This is a consequence of this volume being the first for nonspecialists to present expositions on many of the themes addressed. We have incorporated these lists to help any specialists or advanced students. Also, as scholars, we would feel that we had been remiss if we had hidden all links between this volume and relevant works.

We are indebted to a number of people who have generously assisted in the preparation of this book. Ros Raistrick deserves our special thanks for her skilful and efficient typing of numerous drafts of the chapters; it is always a pleasure to work with her. Despite their busy schedules, Jane Arthur, John Bally, Alan Bridle, Michael Burton, Simon Jeffries, Jon Moores, Bob O'Dell, Mansukh Patel, Julian Pittard, Bo Reipurth, Hugo Schwarz, Ian Stevens, Alan Watson, Jonathan Williams, and Karen Wills each took time to suggest reading or provide visual material that has helped make the book more attractive and informative.

We hope that readers will learn from and enjoy the volume.

CONTENTS

Plates follow page 76.

BLOWING BUBBLES IN THE COSMOS

1

THE FIRST DISCOVERIES
OF ASTRONOMICAL WINDS

Often the mistakes we make in the interpretation of observed phenomena lead to fascinating creations. In Scotland the aurorae were called "The Merry Dancers," whereas people of ancient China may have associated them with "flying dragons." Some Scandinavians thought aurorae to be the ghosts of dead maidens, and in Greenland the indigenous people believed them to be the spirits of stillborn children playing ball. The development of a crude understanding of the aurorae during the 19th century and first decades of the 20th century resulted in the realization that the Sun ejects material reaching the Earth at speeds of many hundreds and sometimes thousands of kilometers per second. Often the material causing a spectacular aurora is ejected in a pulse rather than continuously.

In the 1950s the existence of a fairly continuous solar wind, rather than only impulsive ejections, was firmly established observationally and explained theoretically. The winds of some other classes of stars were discovered observationally before or at roughly the same time that the nature of the continuous component of the Sun's mass loss was first understood.

1.1 The Aurorae

Optical observations of aurorae in March 1716 over large areas of Europe led many academics of that time to consider them to be a new phenomenon.

However, in 1760, the Norwegian historian Schøning argued that the Vikings had reported the sighting of aurorae much earlier. One of their annals, "The King's Mirror," written in about 1230, contains imaginative explanations of the origin of aurorae; the theories include speculations about a big fire encircling the Earth, sunlight stored in ice and later released, and reflection of sunlight by airborne snow crystals.

References to even more ancient sightings exist. Aristotle mentioned glowing clouds in his *Meteorologica*, and in A.D. 37 it was probably an aurora that reddened the sky behind the port of Ostia. The Emperor Tiberius dispatched a garrison to aid citizens whose town he mistakenly believed to be burning. In 1526 Beece, the first Principal of King's College, Aberdeen University, referred to what were most likely a number of auroral sightings in Scotland as early as A.D. 93. Chinese and Korean records cite a probable aurora as long ago as A.D. 1141. The English medieval works *The Anglo-Saxon Chronicle* and *Holinshed's Chronicle* reported auroral displays.

Tycho Brahe observed many tens of aurorae from Uraniburg in the late 16th century. However, there was a dearth of auroral sightings in the 17th century. As will be mentioned later, variations in the properties of the Sun influence the production of aurorae, and there is evidence that the number of sunspots during this period was lower than usual. Hence, in the 17th century there were probably fewer auroral displays in Europe to observe than usual.

Sir Edmund Halley had known of aurorae but had never seen one before 1716. When the March display of that year occurred, he recorded his observations. He possessed what was, for that age, a detailed knowledge of the Earth's magnetic field and correctly surmised that the arched structure of the emission implied a relationship between it and the geometry of the magnetic field.

Aurorae are sometimes classified by their structures. A *band display* extends hundreds to thousands of kilometers horizontally but is only a few tenths of a kilometer to roughly 10 kilometers "thick." A band typically ranges many tens to several hundreds of kilometers upward from a lower border at an altitude near 100 kilometers. Bands with only a bit of curvature are called *arcs*, but other bands are very serpentine in structure. A *patch* or *surface* aurora is an essentially isolated, irregularly shaped region of emission, whereas a *veil* aurora covers a major part of the sky. Terms used to identify auroral internal structure include *homogeneous*, *rayed*, *striated*, and *diffuse*, and are self-explanatory. Plate 1 shows an aurora.

An aurora can be faint or invisible to the naked eye or as bright as the Moon. Aurorae that are readily observable far from the Earth's magnetic poles are called *great aurorae*. Most of the optical emission of aurorae is in a few narrow wavelength bands. Discrete red and green features are emitted by atomic oxygen. A number of blue features arise from the molecular nitrogen ion N_2^+, and red features of the nitrogen molecule N_2 are also seen.

The structure, brightness, and color of an aurora vary in time. In some cases, large-scale variations in the emission give the impression of rapid movement back and forth across much of the sky in seconds. An event may continue for several hours.

Carl Friedrich Gauss, in his work on the structure of the Earth's magnetic field, recognized that electric currents flowing in the Earth's atmosphere may produce a component of the Earth's field. He pointed to aurorae as phenomena possibly caused by electric currents. Many decades later in 1896, Kristian Birkeland showed convincingly that auroral emissions result from the collisions of energetic charged particles with atmospheric gas. This Norwegian scientist working in Oslo created conditions similar to those in the upper atmosphere in a cathode ray tube, a precursor of the television picture tube. He inserted a magnet to show that the energetic cathode rays, now more commonly called electrons, produced emission primarily in rings about its two poles just as the aurorae are concentrated about the Earth's magnetic poles.

The idea that the Sun is the source of particles interacting with the upper atmosphere to produce aurorae dates back to 1733, when the French astronomer J. J. de Mairan proposed that aurorae are caused by the interaction of the Sun's and Earth's atmospheres. It received potential support from observations made in 1859 by R. C. Carrington at the Kew Observatory. On 1 September at 11:18 Greenwich Mean Time he saw two patches of very intense white light appear in a large sunspot. By 11:23 the last traces of these patches had faded, returning the sunspot's appearance to what it had been before. Eighteen hours later a long and violent magnetic storm began. At the onset of this magnetic storm a remarkable aurora was seen over much of the world, including locations as far from the Earth's magnetic poles as Puerto Rico at a latitude of 18° N. Even so, the general belief at the time was that the appearance of the great aurora following the solar flare observed by Carrington was a mere coincidence.

The causal relationship between solar activity and the frequency of great aurorae was established by a statistical analysis of data covering events over a period of roughly 100 years, rather than by the observation of a single giant flare and a resulting terrestrial display. In 1844 H. Schwabe discovered the existence of a solar cycle, during which the number of sunspots decreases and then increases again over about an 11-year period (though there have been intervals, including the famous Maunder minimum of the late 17th century, when this simple periodic behaviour did not obtain). S. Tromholt did some of the early work on establishing the correlation between the frequency of great aurorae and sunspot number. He collected together data that had been gathered between 1780 and 1877. In fact, most aurorae are not great aurorae, and the relationship to solar activity of the frequency of most aurorae differs from the relationship found for great aurorae. Though induced by the solar outflow, the other aurorae are generated by a mechanism differing from that triggering the great aurorae.

The high speed that material from the Sun must possess to cause an aurora 18 hours after expulsion might have seemed implausible to some in 1859. We can make a rough estimate of the speed of the bulk of the particles ejected in the solar flare observed by Carrington if we assume that they induced the brilliant aurora beginning 18 hours later. The distance to the Sun is known to be about 1.5×10^8 km (as described more fully in section 2.1 the notation 10^8 indicates 1 followed by 8 zeros, which is 100 million; km stands for kilometers). An hour contains 3,600 s (s is for seconds). Thus, the speed of solar flare particles was about 2000 km s^{-1} where km s^{-1} means kilometers per second. For comparison, Concorde flying at Mach 2 and a space ship orbiting the Earth in 90 minutes travel at speeds of about 0.3 km s^{-1} and 10 km s^{-1}; the Earth revolves about the Sun only a bit faster than the space ship moves with respect to the Earth.

Not all aurorae are associated with huge solar flares, a fact pointing to the existence of a more continuous solar outflow, which we address in the next section. Though he or she did not realize so, a prescientific human looking at an aurora was using the Earth's magnetic field and atmosphere as a huge detector of the solar wind. We will return to the theme of the atmosphere as a huge detector in section 9.5 when describing ultrahigh energy cosmic rays.

1.2 The Continuous Solar Wind

From the early 1930s to the early 1950s, S. Chapman and V. C. A. Ferraro made concerted efforts to develop a theory of aurorae and terrestrial magnetic storms. In doing so they addressed the nature of the solar outflow. They, like F. A. Lindemann before them, supposed it to consist of an electrically neutral stream of particles in what is now called a plasma. A plasma contains positively charged particles and negatively charged particles that are free. By a free particle we mean a particle that is not bound to any other single particle, even if forces prevent the free particle from moving far from the multitude of the particles. Chapman and Ferraro believed correctly that the plasma they were considering is composed primarily of protons and electrons. At that time only fairly limited data constrained the nature of the stream. That data included observations of the zodiacal light, which is sunlight scattered by electrons in interplanetary space. The zodiacal light data provided only upper bounds on the electron densities.

The concept of a continuous flow of plasma from the Sun originated from the use of comets as probes of the interplanetary medium. In 1951, Ludwig

Biermann postulated that the observed velocities of ionized molecules in some cometary tails are due to acceleration arising from interactions with a continuous solar outflow.

A continuous flow that does not vary with time is referred to as a *steady* flow. In the late 1950s, Eugene Parker of the University of Chicago made a profound contribution to astrophysics. He showed that a flow that is moving radially outward in a hot region (such as the corona of the Sun) into one of much lower pressure can be steady only if it, at large distances from its source, moves faster than the speed of sound. We return to a fuller discussion of Parker's solar wind theory in section 4.2.

1.3 Early Discoveries of the Winds of Stars Other Than the Sun

The year 1929 was pivotal for the study of astronomical winds. In that year a key paper by S. R. Pike on the nature of classical novae appeared. *Nova* is the Latin word meaning "new." Novae brighten considerably for short periods of time and so were often mistaken in the past for new objects. Pike explained the observed variations with time of the luminosities of novae. The explanation is based on the assumption that, following a nova outburst, a hot obscuring wind expands away from the central object. Section 11.3 contains a fuller treatment of novae. Here we simply note that an understanding of winds in objects much more distant than the Sun was developing before the existence of the continuous solar wind was established.

A seminal paper by C. S. Beals was also published in 1929. In it he reported the discovery of winds from what we now know are highly evolved stars, each having a mass tens of times that of the Sun. Most of the stars studied by Beals are carbon-rich Wolf-Rayet stars, which are described in section 3.3.2. Two other stars observed by Beals are eta Carina and P Cygni, each of which is a complex source, difficult to classify. P Cygni gives its name to the characteristic shape or *profile* of the plot of a spectral feature's brightness versus wavelength that often arises due to the presence of a wind. Plate 2 is an image of eta Carina, and panel (b) of figure 1-1 shows a P Cygni profile. To understand that figure and how Beals concluded that the stars he observed possess winds, we must first consider a few elementary aspects of spectroscopy.

Some electrons have enough energy that they are free to escape from the vicinity of any atom, molecule, or ion. Such electrons can possess any energy in a continuous range from zero up to infinite energy. Radiative transitions between different *continuum states* (i.e., any states in which the electron is free and has an energy in the continuous range) occur. The transitions involve the emission of radiation if the electron loses energy and absorption if it gains energy. Transitions between different continuum states cause a hot gas to emit or absorb radiation over a continuous range of frequencies

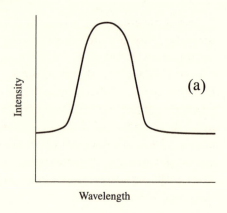

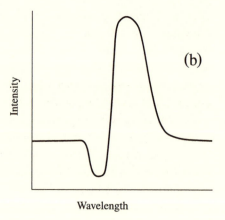

FIGURE 1-1   The Nature of a P Cygni Profile. Panel (a) shows the brightness or intensity of light, as a function of wavelength, emitted in a region of the spectrum where there is a line, and observed in the absence of any absorption taking place between the source and the observer. Panel (b) shows a P Cygni profile arising from a combination of emission followed by absorption taking place in a wind.

or wavelengths (cf. section 8.4). When radiation is emitted over a continuous range of frequencies or wavelengths, it is described as *continuum* radiation or emission. The frequency of the radiation involved in a transition is proportional to the energy change of the electron; the wavelength is the speed of light divided by the frequency.

The spectrum emitted by a hot gas is not necessarily smoothly distributed in wavelength (or frequency). In many situations much of the radiation is emitted in transitions between discrete states of atoms and molecules within the gas. In quantum theory, any state in which all particles in the system are bound has a discrete energy. Transitions between bound states release or absorb discrete amounts of energy. Each type of atom or molecule has its own peculiar set of discrete transitions, and a particular wavelength and corresponding frequency are associated with each such transition. For instance, light with a wavelength of 121.6 nm (nm is the abbreviation for nanometers) is emitted when an electron undergoes a transition from its first excited state to its ground state in an atom of hydrogen.

The spectrum of a star comprises a continuum of emission over all frequencies or wavelengths in a continuous range with narrow (in frequency or wavelength) emission or absorption features superposed on the continuum. Panel (a) of figure 1-1 shows the brightness, as a function of wavelength, of a typical emission feature from a star with no appreciable wind. At wavelengths or frequencies away from the discrete feature the brightness of the emission is at the continuum level. The brightness increases over a narrow range of wavelengths or frequencies. The shape or profile of the feature in the brightness versus wavelength (or frequency) plot is symmetric.

The P Cygni profile shown in panel (b) of figure 1-1 is not symmetric. At lower frequencies or higher wavelengths the profile is similar to the profile shown in panel (a). However, at higher frequencies or lower wavelengths not far from the center of the profile the brightness dips below the continuum level of brightness. As argued below, absorption of radiation by a wind causes the dip.

A star emits because it is hot. However, as depicted in figure 1-2, cooler gas in front of a star usually absorbs some of the stellar emission. The absorption is a consequence of the fact that under most circumstances a greater fraction of particles are in lower states of excitation in cooler gas than in

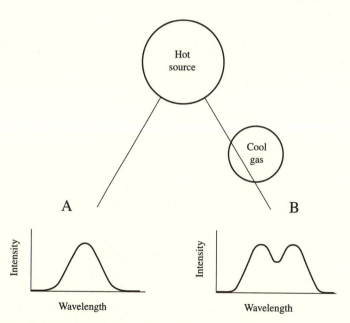

FIGURE 1-2   Absorption in Cool Foreground Gas. An observer at point A has a direct view of the hot source and sees unobscured emission from a discrete transition. An observer at point B measures a different line profile because some of the light emitted by the hot source is absorbed in cooler foreground gas.

warmer gas. Thus, light at the right wavelengths (or frequencies) can more readily induce excitations in the atoms and molecules in cooler gas than in hotter gas. When light induces excitation, light is absorbed. Therefore, the observed brightness of a star is reduced if cooler gas lies between it and the observer.

Motions in the emitting gas or in the absorbing gas can be studied through the analysis of the way that the intensity or brightness varies with wavelength. This is because motions shift the wavelength of light. The shift in wavelength is due to the Doppler effect, which is commonly noticed when an ambulance passes. As the ambulance moves toward someone the pitch of its siren sounds higher than when the ambulance moves away. The same effect can influence the appearance of a source of light. When the source moves toward someone the light appears to have a higher frequency and shorter wavelength than when the source is at rest relative to the observer. When the source moves away the light appears to have a lower frequency and longer wavelength. Because blue light has a higher frequency and shorter wavelength than red light does, light from a source moving toward an observer is said to be *blue shifted*, whereas light from a receding source is said to be *red shifted*.

We are now prepared to consider the information contained in a P Cygni profile. Such a profile is formed by emission from hot gas moving at a range of velocities (some of which are directed away from the observer and some of which are directed toward the observer) and absorption in *cooler* foreground gas moving *toward* the observer. In the absence of the cool foreground the observed profile would appear like that shown in panel (a) of figure 1-1. The presence of the absorption is what indicates the existence of foreground gas that is cooler than gas behind it. The fact that the absorption is concentrated on the short wavelength or *blue* portion of the line shows that it arises in gas moving toward the observer.

Figure 1-3 illustrates how an expanding wind gives rise to the formation of a line with a P Cygni profile. Considerations of emission and absorption occurring at the various points marked A, A', B, B', and C show that emission in gas on the far side of the star moving away from the observer experiences the least absorption in the direction of the observer; hence, the long wavelength or red part of the line profile is nearly the same as in a pure emission profile. In contrast, emission from hot gas moving toward the observer is attenuated by absorption. In 1929 Beals was able to infer from the P Cygni profiles he discovered that the winds of the carbon-rich Wolf-Rayet stars have speeds of about 2000 km s^{-1}.

In 1956 A. J. Deutsch published an analysis of P Cygni profiles detected toward M supergiants. Supergiants are described in section 3.3.2. The M indicates that the stars' atmospheres have more oxygen than carbon. Deutsch showed that they possess much slower winds than the Wolf-Rayet stars. The speeds of their winds are around 10 km s^{-1}.

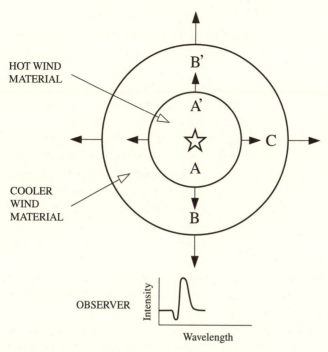

FIGURE 1-3   The Formation of a P Cygni Profile in an Expanding Wind. The temperature of gas decreases with distance from the central star; for simplicity, we have assumed that gas is either hot or cool rather than consider a realistic temperature profile. Light emitted at point A is absorbed in cooler material at point B. However, light emitted at points A′ and B′ is too red-shifted for it to be absorbed in the gas at point B. Because gas at A′ is hotter than gas at B′, it does not absorb radiation emitted at B′. Since gas at point C is cooler than gas at point A it does not emit as copiously, but because its velocity differs from that at point B more than that at point A does, emission from point C suffers less absorption in the direction of the observer than emission at point A.

As mentioned above, the existence of a continuous solar wind was not established until the 1950s. As related in section 3.2, the Sun is currently a main-sequence star and will spend by far most of its life as one. The stars observed by Beals and Deutsch are much more massive than the Sun and have left the main-sequence. In 1957 and 1958, papers by the British scientist Robert Wilson presented results of observations of spectral lines formed in main-sequence O and B stars, which are tens of times more massive than the Sun. He discovered winds with speeds of around 1000 km s^{-1}, and inferred that for such stars, the rate of mass loss (mass carried away by a wind per unit time) is correlated with the radiative luminosity (power) of the star.

This correlation gives insight into the process driving mass loss in these stars. However, a detailed theoretical model of winds driven by radiation pressure (a mechanism addressed more fully in section 7.2) did not appear until 1970, when the important paper by Leon Lucy and Phillip Solomon was published. Their work was stimulated in part by the observation, with a rocket-borne instrument, of P Cygni profiles in far ultraviolet lines (with wavelengths of about 91.2 to 121.6 nm) formed in main-sequences O and B stars, reported by Donald Morton in 1967 (cf. section 7.2).

The growth of ultraviolet spectral astronomy in the 1970s was marked especially by the launches of the very successful Copernicus mission (led by Lyman Spitzer of Princeton University) and of the remarkable International Ultraviolet Explorer satellite (proposed by Robert Wilson, whose discovery of O and B star winds is mentioned above). Those facilities made possible a huge increase in the understanding of a wide variety of astronomical winds. Ultraviolet spectroscopy's main advantage over optical spectral techniques in investigations of many types of winds is that the strongest spectral features in these winds are at ultraviolet wavelengths. The winds' temperatures are in ranges favoring the formation of such features.

We will show later that explosions can be viewed as powerful winds of short duration. We treat the types of bubbles blown by winds and explosions in fairly similar ways. Data obtained over broad spectral ranges have been important for progress in the field of winds, bubbles, and explosions in astrophysics.

SELECTED REFERENCES

Bath, G. T., and Harkness, R. P. "Eruptions of Novae." In *Classical Novae*, ed. M. F. Bode and A. Evans, 61–72. Chichester: John Wiley & Sons, 1989.

Bone, N. *The Aurora–Sun–Earth Interactions*. 2nd ed. Chichester: John Wiley & Sons, 1996.

Brekke, A. *Physics of the Upper Polar Atmosphere*. Chichester: John Wiley & Sons, 1997.

Conti, P. S. "Mass Loss in Early-Type Stars." *Annual Review of Astronomy and Astrophysics* 16 (1978): 371–392.

Davis, T. N. "The Aurora." In *Introduction to Space Science*, ed. W. N. Hess, 205–250. New York: Gordon and Breach Science Publishers, 1965.

Kondo, Y., Wamstecker, W., Boggess, A., Grewing, M., de Jager, C., Lane, A. L., Linsky, J. L., and Wilson, R., eds. *Exploring the Universe with the IUE Satellite*. Dordrecht: D. Reidel Publishing Company, 1987.

Kuhi, L. V. "Wolf-Rayet Stars in the Old Days." In *Wolf-Rayet Stars and Interrelations with Other Massive Stars in Galaxies—Proceedings of the 143rd Symposium of the International Astronomical Union*, ed. K. A. van der Hucht and B. Hidayat, 11–15. Dordrecht: Kluwer Academic Publishers, 1991.

Morton, D. C. "The Far Ultraviolet Spectrum of Six Stars in Orion." *Astrophysical Journal* 147 (1967): 1017–1024.

Ness, N. F. "The Interplanetary Medium." In *Introduction to Space Science*, ed. W. N. Hess, 323–346. New York: Gordon and Breach Science Publishers, 1965.

Newell, P. T., Ming, C. I., and Wang, S. "Relation to Solar Activity of Intense Aurorae in Sunlight and Darkness." *Nature* 393 (1998): 342–344.

Strömer, C. *The Polar Aurora*. Oxford: Clarendon Press, 1955.

Willis, A. J., and Hartquist, T. W. "Sir Robert Wilson, C.B.E., F.R.S." *Astrophysics and Space Science* 237 (1996): 3–9.

2

THE MAGNITUDES OF
ASTRONOMICAL QUANTITIES

Before continuing on the physics of astronomical winds, bubbles, and explosions, we look briefly at the ranges of sizes and properties of astronomical sources, and in the next chapter at the evolution of stars, which are common sources of winds.

Our comprehension of the magnitudes of measurable quantities is largely governed by our everyday experiences. A mile-long stroll is a "short" trip; a flight to Australia from England is a "long" trip. We shiver in a temperature of 0°C; we bask in a temperature of 30°C. A minute goes quickly; a half century passes slowly (at least to begin with).

In astronomy we usually meet quantities having magnitudes that lie completely outside the range of our experience. In this chapter we set the scene by discussing some of the more obviously important astronomical parameters such as mass, distance (or length), and time, as well as density and temperature. However, we do not address the ways in which they are measured.

The reader may be unfamiliar with many of the types of objects for which the chapter's tables contain entries and should refer back to those tables when reading further.

2.1 Big and Small Numbers—Notation

Large numbers are associated with cosmological distances, the masses of black holes at the centers of active galaxies, the temperatures at the centers of stars, and the speeds of material ejected in supernova explosions. Processes on the atomic, molecular, and nuclear scales govern the energy generation and emission of stars; the masses and lengths relevant to single atoms, molecules, and nuclei are small numbers when expressed in everyday units.

In section 1.1 we introduced a notation for writing big numbers when examining the speeds of ejecta from the Sun. In that notation one writes 100 as 10^2 and 1812 as 1.812×10^3. The index on 10 tells us how many times to multiply by ten. We employ a similar notation when considering small numbers. Using it one writes 0.001 as 10^{-3} and 0.0035 as 3.5×10^{-3}. The minus sign in the index of the 10 signifies that one should divide by 10 the number of times indicated by the index.

2.2 Mass

The mass of a body is simply a measure of the amount of material contained in it. The basic building block in the Universe is the hydrogen atom. (Its mass is 1.67×10^{-27} kg; kg stands for kilograms). Sometimes the mass of a hydrogen atom is not a very convenient unit. For example, the Sun contains about 10^{57} hydrogen atoms (more properly hydrogen nuclei, as most of the hydrogen is ionized), and the observable Universe as a whole contains about 10^{78} hydrogen nuclei. Astronomers generally find the mass of the Sun to be a convenient unit. It is called a *solar mass*, is symbolized by $M_\odot$, and is about 2×10^{30} kg. The expression of the mass of a system in solar masses has the additional advantage that a "feel" for the approximate number of stars in the system (if on average each has a mass of about 1 $M_\odot$) is given. Table 2-1 lists some important masses.

2.3 Distances and Lengths

Astronomical distances and lengths are vast, making our usual units of measurement singularly inappropriate. For example, a typical galaxy has a ra-

TABLE 2-1
Important Masses in Astronomy

Mass of H atom	1.67×10^{-27} kg	10^{-57} $M_\odot$
Mass of Earth		3×10^{-6} $M_\odot$
Mass of planetary nebula		0.5 $M_\odot$
Mass of Sun		1 $M_\odot$
Mass of bright young star		50 $M_\odot$
Mass of interstellar cloud		10^4 $M_\odot$
Mass of the Milky Way		8×10^{10} $M_\odot$
Mass of the Visible Universe		10^{21} $M_\odot$

dius of roughly 10^{18} kilometers. Rather analogously to the way our everyday units are tied to lengths and distances commonly encountered (e.g. the meter and the nautical mile), astronomers define units for lengths and distances to suit circumstances. Thus, the radii of stars are usually expressed in terms of the radius of the Sun (1 $R_\odot$ = 6.96 × 10^5 km); the distances of planets or asteroids from the Sun (or indeed other stars) are expressed in terms of the AU or Astronomical Unit (1 AU = mean Earth-Sun separation = 1.5 × 10^8 km).

However, even such large units are themselves impractical for many purposes. The Sun, for example, is located at a distance of approximately 2.7 × 10^{17} km from the center of the Milky Way; to express this as 1.8 × 10^9 AU is hardly more convenient. A useful unit is the light year. This is the distance travelled in one year (3.16 × 10^7 s) by light, which moves at the constant speed of about 300,000 km s^{-1}. Hence 1 light year is approximately equal to 9.5 × 10^{12} km. This unit also has the advantage (by definition!) that distances expressed in this way tell us how long in the past light, that we now receive, set off toward us. It also enables us to properly time sequence a set of events. For example, event A may (to us) precede event B, but if the light travel time to us from the location of B is longer than that from the location of A, then physically B may have preceded A.

Another commonly used unit of length is the parsec, which is about 3.08 light years. A star at a distance of 1 parsec would appear to shift its position on the sky by 1 arcsecond, which is 1 degree of arc divided by 3,600, as an observer moved 1 AU along a straight path. Astronomers measured such shifts in the apparent positions of stars, caused by the Earth's motion around the Sun, to make the first determinations of distances to stars other than the Sun. The biggest unit of length commonly employed by astronomers is the megaparsec, which is 1 million parsecs. Many galaxies belong to clusters of galaxies having extents of roughly a megaparsec. Just as both the mile and the kilometer are suitable units for giving distances between towns, the parsec and the light year are appropriate units for expressing the sizes of or distances between many types of astronomical objects. However, we use light years rather than parsecs throughout most of the book.

Table 2-2 lists some important distances and lengths.

2.4 Astronomical Timescales

The timescales between events, or over which events can take place, are of fundamental importance to astronomers. For example as we discuss in chapter 3, stars do not maintain forever the physical structure with which they are born. Rather, they *evolve*. The timescale for evolution depends mainly on the ratio of the amount of the particular nuclear fuel a star is burning

TABLE 2-2
Important Distances and Length Scales

Mean Earth–Moon distance	3.844×10^3 km	0.02 light seconds
Mean solar radius	6.96×10^5 km	2.3 light seconds
Mean Earth–Sun distance	1.495×10^8 km	8.3 light minutes
Planetary nebula radius		0.3 light years
Distance to nearest star (α Centauri)		4.3 light years
Wind blown bubble radius		15 light years
Molecular cloud radius		50 light years
Old supernova remnant radius		300 light years
Sun–Orion Nebula distance		1,400 light years
Sun–Galactic Center distance		27,000 light years
Galaxy radius		50,000 light years
Milky Way–nearest galaxy distance		3×10^5 light years
Milky Way–Virgo cluster distance		5×10^7 light years
Milky Way–quasi-stellar object distance		10×10^9 light years
Radius of the Visible Universe		15×10^9 light years

to the mass consumption rate of the burning process. A star similar to the Sun has fuel to power roughly 10 billion years of hydrogen burning. (We use billion to indicate 1,000 million rather than 1 million million as the British did until recently.) A massive blue giant star that contains, say, 50 times the amount of hydrogen as that in the Sun, will burn for only about 10 million years because it is much more profligate in its mass consumption.

Even longer and, indeed, shorter timescales are relevant. The longest time in astronomy is the present age of the Universe (the time elapsed since the Big Bang). A reasonable estimate for this is 15 billion years. The shortest time of astronomical significance is the Planck time (10^{-43} s), which was the age of the Big Bang when gravitational and quantum mechanical effects were of comparable importance. More relevant to the discussion of this book is the time an excited hydrogen atom stays in a high-energy level before decaying to a lower energy level and producing a photon that eventually carries information to waiting astronomers. This time is typically about one hundredth of a millionth (10^{-8}) of a second.

There are some specialized timescales that are used in connection with some of the dynamical events described later in the book. One is the *expansion time*, which is the time required for a bubble to expand to some particular size or radius. An adequate estimate of this time can be obtained by the division of the bubble size by its current expansion speed. A related timescale is the *flow time*, the time taken for material to move through a flow region of a certain length. This time is reasonably approximated by the division of

TABLE 2-3
Important Timescales

Lifetime of an atomic energy level	10^{-8} second
Duration of a nova	10 weeks
Duration of a supernova	1 year
Lifetime of a planetary nebula	10^4 years
Expansion time of a wind-blown bubble	10^5 years
Expansion time of a supernova remnant	10^6 years
Flow time through a galactic superwind	10^7 years
Rotation period of a galaxy	10^8 years
Age of Earth	4.5×10^9 years
Age of Sun	5×10^9 years
Age of Milky Way	12×10^9 years
Age of Visible Universe	15×10^9 years

that length by the typical speed of the flow. Table 2-3 lists a collection of important timescales.

2.5 Astronomical Densities

Astronomers use two forms of density. One is the usual mass density, which is just the mass per unit volume. For some purposes though, it is more convenient and informative to use the number density. This is defined as the number of particles per unit volume. The particles could be stars or even galaxies, but we usually employ number density to mean the number of nuclei per unit volume. Counting the number of nuclei is convenient because we need not specify whether the material is solid, atomic, molecular, or ionized.

The units used depend on circumstances. If we are, for example, describing the mass density in a star, then units of g cm^{-3} (which indicates grams per cubic centimeter) are usual. (It is interesting to note that the mean mass density in the Sun is about equal to that of water.) On a much grander scale, cosmologists may use units of solar masses per cubic megaparsec. A cubic megaparsec is about 3×10^{67} cubic meters.

In general, matter in diffuse astronomical objects such as nebulae or ejected stellar envelopes is so rarefied that it makes more sense to use number densities rather than mass densities. So as an example, if we took all the gas in the disk of a spiral galaxy such as our own and worked out the average mass density, we would find it to be about 10^{-24} g cm^{-3}, corresponding to about one particle (mostly hydrogen) per cubic centimeter. Clearly the use of the number density provides a better "feel" for the nature of conditions.

Finally, there are rather less obvious circumstances in which astronomers find number densities more convenient. An important question having a strong bearing on the ways in which flows behave concerns just how quickly gas loses energy by radiation. Astrophysicists assess the importance of energy loss by comparing how quickly the gas loses energy to, say, the expansion time for a bubble (cf. section 2.4). Obviously, energy losses would affect the expansion of a bubble if the energy loss time (usually called the cooling time) were significantly less than the expansion time. Energy loss times are usually proportional to the inverse of the gas density. It is easier for many of us to comprehend quickly that the cooling time for a gas of number density 1 cm^{-3} is ten times less than for a gas of number density 0.1 cm^{-3}, rather than think about the quantities of 10^{-24} g cm^{-3} or 10^{-25} g cm^{-3}. Table 2-4 gives the densities of a number of astronomical sources.

2.6 Temperature

Throughout this volume temperatures are measured on the Kelvin scale, the scale on which the temperature is zero (or absolute zero) when all matter is in its lowest energy state. On this scale, each degree change corresponds to a change of one degree centigrade (or Celsius). However, absolute zero corresponds to a temperature of approximately –273 degrees centigrade (or –273°C). We write temperatures on the Kelvin scale as, e.g., 300 K (which is about room temperature).

From the physicist's point of view, the definition of temperature is rather general. It is that property of a system that determines whether there is a net flow of heat out of this system, or into this system, when placed in contact with some other system. There are two particularly common ways in which the concept of temperature can be related to some physical property of a system.

TABLE 2-4
Important Densities in Astronomy

Average density of stars in the Milky Way	1 M$_\odot$ pc^{-3}	1 particle (star)
Average density of the Universe	10^9 M$_\odot$ Mpc^{-3}	0.1 particles cm^{-3}
Average density of hot gas in a bubble	10^{-26} g cm^{-3}	10^{-2} particles cm^{-3}
Average density of gas in the Milky Way	10^{-24} g cm^{-3}	1 particle cm^{-3}
Average density of gas in a planetary nebula	10^{-20} g cm^{-3}	10^4 particles cm^{-3}
Average density of the broad line region of an AGN*	10^{-14} g cm^{-3}	10^{10} particles cm^{-3}
Average density of the Sun	1 g cm^{-3}	10^{24} particles cm^{-3}
Average density of a white dwarf Star	10^5 g cm^{-3}	10^{29} particles cm^{-3}
Average density of a neutron star	10^{15} g cm^{-3}	10^{39} particles cm^{-3}

*AGN indicates active galactic nucleus; see chapter 10.

The temperature is the measure of the kinetic energy per particle associated with the random motions of particles (be they atoms, ions, or molecules) within a gas, plasma, liquid, or solid. The average kinetic energy per particle due to random motion in an atomic gas is proportional to the temperature (T). Since by definition, the kinetic energy of a (nonrelativistic) particle is just one half the particle mass times the square of its velocity, then T is a measure of the speed of the particles' random motions. In many circumstances, even if a gas contains particles of different masses, all particle types have the same kinetic energy per particle associated with the random motions caused by the sharing out of energy over many collisions (this is called *equipartion of energy*). So for example, if a gas contains protons and oxygen ions, the latter have average random speeds about one quarter those of the protons, because they are about a factor of 16 more massive.

There are various ways of raising the temperature of gas under astronomical circumstances. One way is to pump in energy carried by radiation. A common mechanism leading to heating due to radiation is the photoionization of hydrogen (i.e., ionization by light). In this process, electrons are ripped away from hydrogen atoms by energetic photons. The electrons are given kinetic energy, which they share out by collisions with all other particles present. Another way is by means of a compressive shock wave (cf. section 4.1), a phenomenon generated by aircraft flying at speeds greater than the sound speed in air. The temperature to which gas is raised is proportional to the product of the mean mass per particle of the gas that is shocked and the square of the speed of the shock wave relative to the gas it is overrunning. As a guide, atomic hydrogen gas overrun by a shock moving at 1000 km s^{-1} is shocked to a temperature of about 10 million degrees Kelvin. Clearly, shocks with speeds of 100 km s^{-1} and 10 km s^{-1} would raise the gas temperature to about 100,000 and 1,000 degrees Kelvin, respectively.

There are other ways of heating gas to high temperatures. The appearance of the solar corona is a spectacular sight revealed to the naked eye only during solar eclipses. The solar corona is very hot (about 1 million K), far hotter than the temperature of the Sun's surface layers (about 6,000 K). The conversion of energy stored in magnetic fields to heat probably causes the high coronal temperature (cf. section 7.1).

The concept of temperature can also be used to describe the properties of a radiation field (e.g., that emitted by a star or that in the central regions of an active galaxy). At its very simplest level, we could say that the temperature of the radiation field, T_R, is a measure of the energy of a typical photon in the radiation field. The energy of a photon is inversely proportional to the wavelength, so that a red photon has an energy less than that of a blue photon. Since the photon energy is proportional to T_R, obviously the higher the temperature of the radiation field the bluer the photons emitted. This is precisely why hot stars are blue and cool stars are red.

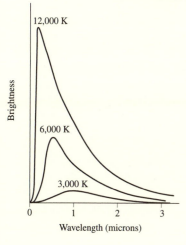

FIGURE 2-1   The Distributions of Brightness (or Intensity) Produced by Black Bodies. Brightness is plotted versus wavelength for different temperatures. The wavelength at maximum brightness systematically decreases as the black body temperature increases.

A more rigorous usage of temperature in connection with radiation fields necessitates the introduction of the concept of a black body radiation field. Such a radiation field has a particular type of distribution of energy with wavelength or frequency, which is shown in figure 2-1. A black body distribution peaks at a wavelength that is proportional to $1/T_R$. The power radiated per unit surface area by a black body is proportional to the fourth power of its temperature; consequently, one square meter of the surface of a star at 6,000 K radiates at a power about 16 times that of one square meter of the surface of a star at 3,000 K.

In Table 2-5 we give a set of temperatures (gas and radiation) for some important astronomical circumstances.

TABLE 2-5
Astronomical Temperatures

Cosmic microwave background	2.7 K
Cool molecular cloud	10 K
Neutral atomic cloud in Milky Way	100 K
Surface of Sun	6,000 K
H II* region	8,000 K
Planetary nebula	12,000 K
Broad emission line cloud in AGN	30,000 K
Surface temperature of young massive star	50,000 K
Planetary nebula nucleus	150,000 K
Solar corona	10^6 K
Hot gas driving a 1000 km s^{-1} wind bubble	10^7 K

*H II indicates ionized hydrogen; see section 6.1.

REFERENCES

Allen, C. W. *Astrophysical Quantities*. 3rd ed. London: Athlone Press, 1973.

Audouze, J., and Israël, G. *The Cambridge Atlas of Astronomy*. Cambridge: Cambridge University Press, 1994.

Rowan-Robinson, M. *The Cosmological Distance Ladder*. New York: W.H. Freeman & Co., 1985.

3

STELLAR EVOLUTION

The physical structures and properties of stars change as they age. In this chapter, we give an overview of the way stars evolve to set the context for our later discussions. The mass is by far the most important single factor determining a star's structure and evolution. A corollary of this statement is that any significant mass loss from a star affects both its structure and subsequent evolution. Much of present research into stellar structure and evolution involves the study of stars whose masses change significantly as they evolve.

3.1 Stellar Structure

Various parameters, such as the mass and the helium-to-hydrogen ratio, affecting a star's structure change as a star evolves. However, they do so on timescales that are long compared to the timescale for the structure to respond to a change in the parameters. For instance, in the Sun nuclear burning is causing a significant change in the helium-to-hydrogen ratio on a timescale of roughly 10 billion years. In contrast, the Sun would take roughly only 24 million years to return to its present structure if some unforeseen near-collision with another star were to disturb the Sun. Some 24 million years is the timescale on which the Sun would have to be collapsing if the release of gravitational energy rather than nuclear processes were responsible for the Sun radiating at its observed power or luminosity. The "relatively" short timescale on which the structure of a star can adjust means that we can often consider a star's interior to be in *static equilibrium*. By *static* we mean that motions are negligible, and by *equilibrium* we mean a state in which the different forces acting on any part of the star's interior add up to zero.

Only one force keeps a star together and that is its gravity (or, more accurately, its self-gravity). Any piece of material inside a star feels the gravitational attraction of all the material lying nearer to the center of the star than itself. This force is proportional to the mass of this interior material and inversely proportional to the square of the distance of the piece of material from the center of the star. For the purposes of calculating the gravitational attraction, the interior material behaves as if it were all located at the center of the star.

Several forces act to disrupt a star. These include the forces associated with thermal gas pressure, the pressure of light inside a star, and centrifugal acceleration if the star rotates. The relative importance of such forces depends very much on circumstances. For simplicity we first consider stars (such as the Sun) in which the most important disruptive force is due to gas pressure.

Stars are very much denser than the gas distributed throughout a galaxy. As mentioned in section 2.5, the average density in the Sun is about the same as the density of water. However, because the interiors of stars are so hot, we can still think of pressure in the same way that we do when thinking about the pressure of air. In an ordinary gas, like that in a room, the gas pressure is proportional to the number density (cf. section 2.5) and to the temperature (cf. section 2.6).

A star like the Sun is in a state of equilibrium when the gravitational and pressure forces balance. In figure 3-1 we look at a small shell of material concentric about the star's center and see how forces balance for that shell. Newton's Second Law and his Theory of Gravity tell us that the gravitational force on that shell is given by the product of G, the gravitational constant,

FIGURE 3-1 The Balance of Forces Inside a Star. The small shell with a mass ΔM experiences a gravitational force ΔMg toward the center of the star. This force is balanced by a small pressure difference ΔP across the shell.

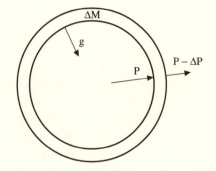

the mass of that shell, and the mass of material interior to that shell divided by the square of the distance of that shell from the star's center, as described above.

When we calculate the pressure force on the shell we have to take into account the fact that the shell has two surfaces; one is nearer the center of the star than the other. Pressure pushes the inner surface outward but pushes the outer surface inward. If the pressure were the same on both surfaces, there would be no net pressure force on the shell to oppose the inward pull of gravity. Since we need the net pressure force to push outward from the star's center, then the pressure on the inner surface must be greater than that on the outer surface, that is, the pressure in the star must vary with position and must decrease away from the star's center. The decrease in the pressure in the Earth's atmosphere with height above the ground is due to the same principles as those governing a star's pressure decrease.

A simple consideration of mechanical force balance is by itself inadequate to enable us to explain self-consistently the interior structure of a star. However, it can be used together with some further input information to make a perfectly sensible estimate of one of the most important physical parameters of a star, namely its internal temperature. We actually have to specify two other parameters, the stellar mass and radius, which can in principle be obtained from astronomical observations. If we use the mass and radius of the Sun, we calculate an interior temperature of several millions of degrees. Temperatures comparable to this and even higher are characteristic of the centers in the vast majority of stars.

In order to proceed further with our description of the interiors of stars, we have to include some additional physics. We need to consider how energy is generated and transported in a star. The Sun's high interior temperature was mentioned above; it is high enough that thermonuclear processes (or nuclear burning) power the Sun. In the simplest form of stellar nuclear burning, four protons fuse together to form a helium nucleus, produce lower mass particles called positrons and neutrinos, and release energy. High temperatures are needed in order for the protons to have enough energy to overcome the mutual repulsion due to their similar positive charges. Energy generated by the reactions is transported outward through the star and finally leaves the surface of the star as radiation.

The temperature in a star decreases with distance from its center. We observe the surface temperature of the Sun to be about 6,000 K and yet we estimate the interior temperature to be several millions of degrees. The laws of thermodynamics tell us that heat travels from high-temperature regions to lower temperature regions. However, by themselves these laws do not tell us how energy is transported. To tell that we need to understand the processes by which such transport occurs.

The three possible processes involve conduction, convection, and radiation. We experience conduction if we grasp a poker inserted into a fire, convection when we stand near a radiator, and radiation when we stand in sunlight. For most stellar interiors, transport by conduction is not important and we therefore say no more about it here.

Convection transfers energy in stars as hot gas rises into cooler regions, gives up some of its energy to the cooler regions, and sinks back down into the star. A circulation pattern is set up exactly in the way a radiator sets up a hot air circulation pattern in a room. Radiation transports energy as photons leave hot regions and move into cooler regions where their energy is converted into heat by absorption.

The transfer of energy by radiation is particularly efficient if the photons (quantum packets of information) can travel easily through the gas inside the star. This is usually the case if the gas is quite hot (more than a few thousand degrees Kelvin, say). However, the transfer of energy by convection is a much more efficient process than the transfer of energy by radiation where photons get absorbed by matter before travelling very far or where large changes in temperature occur over short distances. Both processes can be important in any particular star but, fortunately, they dominate in quite well defined different regions of that star.

3.2 The Results of Stellar Structure Calculations

We will describe the structure and evolution of isolated single stars only, even though a significant proportion (roughly half) of all stars are formed in multiple (e.g., binary) systems. The evolution of stars in multiple systems can be greatly affected by their companions and will not be considered further in this chapter. Some discussion of binary stars is given in sections 11.2 and 11.3.

To begin our summary of stellar evolution, we examine the *Hertzsprung-Russell* (H-R) diagrams of two clusters of stars. The axes of an H-R diagram are the luminosity (or radiative power) of the star and the temperature of its surface. An H-R diagram is constructed from observations of a group of stars. Figure 3-2 shows a schematic diagram for a recently formed cluster; one sees that the luminosity-temperature points representing the stars lie in a band about a rather straight line in the diagram. Stars associated with this band are said to be *main-sequence* stars. Figure 3-3 displays a schematic representation of the H-R diagram that will be appropriate for the stars from the same cluster but after about 10 billion years of evolution; the stars in the main-sequence band are in a small portion of it to the lower right (labelled MS). Other regions contain points corresponding to stars that are no longer "on" the main sequence and are labelled with various designations, such as *red*

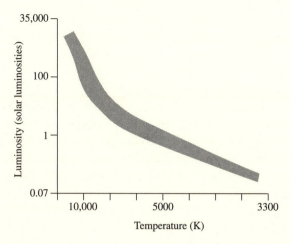

FIGURE 3-2 The Hertzsprung-Russell Diagram for a Recently Formed Cluster of Stars with a Range of Masses. The dark band (the *main sequence*) shows the relationship between luminosity and temperature as the stars burn hydrogen in their cores. The temperature is the surface temperature. The temperature is highest to the left, and the luminosity increases toward the top of the plot.

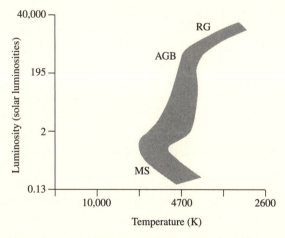

FIGURE 3-3 The Hertzsprung-Russell Diagram of the Same Cluster at a More Advanced Stage of Evolution. Any star that contains about the same or more mass than the Sun has evolved off the main sequence after a time of about 10 billion years. (MS, main sequence; AGB, asymptotic giant branch; RG, red giant; the natures of the types of stars are addressed in sections 3.2 and 3.3).

giant (RG). The theory of stellar evolution successfully explains much of the movement of the point associated with a star on an H-R diagram. The evolution and that movement are consequences of the consumption of the available nuclear fuels.

All stars start their lives burning their basic and most abundant fuel, hydrogen, in their cores. They spend most (typically 90 percent or so) of their lives in this main-sequence phase, which is why stars concentrate in the main-sequence part of the H-R diagrams for many groups of stars. The Sun is in its main-sequence phase of evolution. Stars change little until they are very near to the end of their main-sequence lifetimes. There are two basic interior structures that main-sequence stars have. The mass determines which of these two a particular star has.

Main-sequence stars with masses up to about twice that of the Sun burn their hydrogen to produce helium in a series of nuclear reactions known as the p–p (proton–proton) chain. The rates of the reactions involved do not depend too greatly on the temperature—at least by the standards of nuclear reactions—and the temperature gradient set up deep inside the star is not too severe. As a result, the energy flow demanded is not too great to be carried by radiation, and radiation is what transports energy in the central regions of the star. However, near the outer regions of such a star, the temperature falls to such a level that the gas becomes opaque to radiation and radiative energy transport is so inhibited that convection starts to provide the energy transport. A main-sequence star with a mass less than twice the Sun's therefore has a radiative core and a convective envelope.

Main-sequence stars with higher masses turn hydrogen into helium in a different series of reactions known as the C–N–O chain, which involves the elements carbon, nitrogen, and oxygen. The nitrogen and oxygen are produced in a recycling process that starts with protons interacting with carbon and ends with the creation of helium nuclei and the return of the carbon. The heavier elements are not being burnt; they are catalysts. Unlike the p–p chain, the C–N–O chain is very sensitive to temperature. The temperature sensitivity of the C–N–O chain is important because the more massive a star, the higher is its central temperature. As a result, the central regions of more massive stars generate energy at very high rates, and radiation is simply not efficient enough at transporting the energy outward. Convection is the only mechanism capable of this transport. In the outer layer of a star, no nuclear burning is taking place and there is not a very steep temperature change with position, but the temperature of the stellar material is still high enough for it to be transparent to radiation. Radiation takes over from convection in the outer layers, and such a star therefore has a convective core and a radiative envelope, the inverse of a lower mass star.

The lifetime of a star on the main sequence depends sensitively on its mass for two reasons. First, observational data tell us that the more massive a star, the brighter it shines; in other words, a massive star generates energy at a much greater rate than does a lower mass star. Second, the amount of fuel—hydrogen—available for burning must be proportional to the stellar mass. It is then found that this lifetime is approximately inversely proportional to the square of the mass of the star. Thus, a star that has an initial mass ten times that of the Sun will stay on the main sequence for only one hundredth of the time that the Sun will.

3.3 Stellar Evolution After the Main Sequence

To simplify the following discussion we divide stars into two separate groups, which are divided into subgroups. Our first group comprises low-to-intermediate mass stars, whose masses range roughly between that of the Sun and perhaps eight times its mass. Our second group comprises massive stars whose masses are upward of about eight times that of the Sun. Stars much less massive than the Sun have main-sequence lifetimes greater than the age of the Universe and need not be considered here. To anticipate our later description, we note that perhaps the major difference between the groups is that one of the high-mass stars ends its life with a bang known as a supernova, whereas one of the lower mass stars ends its life with a whimper as it loses material more gradually.

We now introduce the idea of a state of matter rather different from that found in main-sequence stars. Matter in this different state is said to be *degenerate* and it has properties that differ dramatically from those of ordinary gases in which the thermal pressure is proportional to the density and temperature. Although there are many different particles around inside stars (various nuclei, electrons, protons, etc.), we are currently referring to the degeneracy of electrons only. Protons and neutrons become degenerate under far more extreme conditions than are met here. We are familiar with the concept of density in terms of the amount of mass, or the number of particles, per unit volume of space. But particles are always moving and we can also define a new *density* in terms of the number of particles that inhabit some *volume* of velocity space. The great German theoretical physicist Wolfgang Pauli showed in 1925 that particles such as electrons cannot be packed more tightly than a certain limit into a volume made up jointly of real space and velocity space. This limit depends on the gas density and temperature and becomes important if we have a high gas density and a low gas temperature. The higher the density, the greater the gas temperature must be, if the limit is to remain irrelevant. Once conditions cause the limit to be approached, the gas becomes degenerate. Any particles added to the region must have speeds higher than the thermal speed because of Pauli's limit on the amount of packing in the combined real-velocity space. The most significant effect is that the gas pressure becomes totally independent of the gas temperature.

To see the implications of this, imagine a simple experiment in which we take two balloons and fill one with ordinary matter and one with degenerate matter. We surround each balloon with ordinary gas, ensure that their pressure remains constant throughout our experiment, and inject energy at some rate into each balloon. The injection of energy obviously increases the temperature of the material. The pressure in an ordinary gas will increase, the balloon will expand, and the temperature of the expanding material will drop. The balloon containing degenerate matter does not expand because the pressure is independent of temperature. As we consider below, degeneracy limits the extent to which low-mass and intermediate-mass stars release the available nuclear energy. It is not important for nuclear burning in high-mass stars before they explode as supernovae (cf. section 8.1).

3.3.1 Post Main-Sequence Evolution of Low- and Intermediate-Mass Stars

Whatever the mass of a star is, the result of hydrogen burning is the production of helium. The helium sinks to the star's center. Since we use four particles (protons) to produce one helium nucleus, one very important effect of burning is the reduction of the number of particles. (Other particles produced in the reaction escape or are annihilated to form radiation.) The gas pressure is proportional to the number density of particles, and so as burning proceeds, the region where the hydrogen has burnt to helium contracts to maintain the pressure balancing the weight of all the material overlying it. The hydrogen burning is confined to a shell around the helium central sphere. Figure 3-4 depicts the structure of the central region of such a star in this and other phases. While hydrogen burning occurs, the star undergoes a remarkable change in its structure. As the helium core contracts, the outer atmosphere of the star greatly expands to as much as 1,000 times its former extent. The star has become a giant (not to be confused with massive) in spatial extent. There is no simple reason that this expansion takes place. It is generally agreed that several different physical mechanisms are involved, but there is no current consensus as to their relative importance.

The surface temperature of the star in its expanded state is rather low, typically about 4,000 degrees Kelvin. Consequently it appears red in color. Because the star has such a large surface area it radiates at a rate far greater than the star originally did when it was on the main-sequence. This increased energy generation of course has to be provided by the nuclear burning inside the star. These stars are called red giant stars. In this phase stars can lose material in some of the ways described in chapter 7.

What then happens in detail depends rather critically on the mass of the star, though the end effects are essentially the same. We first define low-mass stars as those having masses less than about twice that of the Sun. When

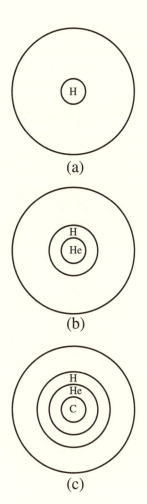

(a)

(b)

(c)

FIGURE 3-4   The Internal Structure at Different Times of a Star with a Mass a Few Times That of the Sun. (a) The star on the main sequence. (b) The star after a few billion years; hydrogen is burning in a shell. (c) The star after about 10 billion years; hydrogen burns in a shell exterior to a helium burning shell.

these stars reach the red giant phase, their cores have contracted so much that they are degenerate. Simultaneously, their central temperatures have increased to the point at which helium starts to burn at the stellar center to form carbon. This is by the mechanism called the *triple-alpha* process, since in it three helium nuclei (i.e., alpha particles) burn to form one carbon nucleus. Heat is generated of course, and the temperature starts to rise. However, because the core is degenerate its pressure does not increase and it does not expand and cool off. The central temperature rises in a few seconds and runs away, leading to a very sudden increase in the rate of helium burning. This is called the helium flash. The energy generation rate increases dramatically, but, perhaps disappointingly, an observer watching the star during this phase would see very little. This is because the increased radiation takes a thousand years or so to get out of the star and the effects of increased energy generation over a time of a few seconds are smoothed out. The core, which is now made up of mainly carbon with some oxygen, contracts again and becomes degenerate. There is still helium burning in a shell around the core. The points corresponding to stars with mainly carbon cores are somewhat to the left (i.e., in the direction of increasing surface temperatures) of those associated with the red giant stars in the H-R diagram depicted in figure 3-3, and these stars are called AGB (asymptotic giant branch) stars.

The intermediate-mass stars behave differently in one main respect. Their helium cores are not degenerate and so when helium burning takes place, they adjust their core pressures by expansion and there is no helium flash. However, the ultimate result is that at the end of helium burning in the cores, they have cores of carbon and some oxygen which, just as in their lower mass cousins, then contract and also become degenerate. They, too, are AGB stars, and helium burning in a shell takes place in each. The subsequent evolution of both low- and intermediate-mass stars is effectively the same.

During the lifetime of an AGB star, complex events involving short flashes of heat produced by runaway thermonuclear burning in helium shells can occur. The flashes are driven by the ignition of hydrogen-burning shells which subsequently provide heat to trigger these runaway events in the helium layers below them. However, by far the most important process taking place is the loss of surface layers of the AGB stars by processes described later in chapter 7. A significant fraction of the mass of the envelope of the star is lost in this way. The ejected mass eventually forms a planetary nebula, one of which is pictured in plate 3. The optical light from such a nebula comes from gas at about 10,000 degrees Kelvin and a number density of about 1,000 particles per cubic centimeter.

A few percent of the star's total mass remains in a thin shell of burning hydrogen, which provides the stellar luminosity. The gradual burning away of the hydrogen shell produces a constant luminosity as the surface temperature increases. Eventually, there is no more hydrogen to burn and the luminosity of the star decreases, because the temperature is not high enough inside the star to ignite the carbon or oxygen there (a much higher ignition temperature than that of hydrogen is required). The central core now contracts, and the material now becomes degenerate. The star becomes a white dwarf, a slowly cooling, spent remnant core. The surface temperature and brightness slowly decrease, and eventually the star is too dim to observe.

Figure 3-5 shows the path of the point in an H-R diagram associated with a star with a mass several times that of the Sun as it evolves. One important caveat to the rather dreary white dwarf evolution outlined above needs mention. The Indian astrophysicist S. Chandrasekhar showed in 1932 that a white dwarf star can have a mass no greater than a certain limit (about 1.4 times the mass of the Sun); he was awarded the Nobel Prize in part for this

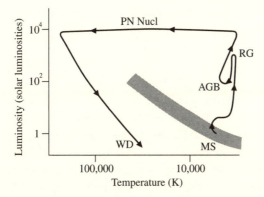

FIGURE 3-5   The Evolutionary Path of a Star with a Mass a Few Times That of the Sun in the H-R Diagram. The star starts on the main-sequence and then evolves over billions of years through the AGB and RG phases to become a planetary nebula nucleus (PN Nucl). It evolves through this phase in a time of thousands of years. Eventually, there are no more sources of nuclear fuel to burn and the star cools to its final white dwarf (WD) stage.

contribution. Stellar cores more massive than this can certainly evolve toward the white dwarf stage, and it is far from clear what actually happens to them. One popular speculation is that some may collapse to form black holes.

3.3.2 Post Main-Sequence Evolution of Massive Stars

Just as for the lower mass stars, hydrogen burning on the main sequence leads to the production of helium, though in the massive star the newly formed helium is mixed through the core by convection. Unlike the cores in the lower mass stars, the helium and carbon cores of massive stars never become degenerate. When any new burning cycle (e.g., helium to carbon) starts, the star can always expand and attain a new stable state as the core pressure increases as a result of that burning. Flashes, such as the helium flash, do not occur and the star goes through a sequence of burning cycles during which heavier and heavier elements are produced. As described in section 8.1, eventually the core burns to iron and then the processes initiating Type II supernova production occur.

There are complex problems regarding the internal physics of massive stars. For example, a proper treatment of the stellar structure at the boundary between the convective core and the outer radiative regions is very difficult and has not yet been achieved. However, when the evolution of these stars is calculated, they show that the stars move redward in the H-R diagram while keeping their energy output rates roughly constant. In other words they, too, swell up just like low-mass stars but without an increase in luminosity. Swollen red high-mass stars are called red supergiant stars. We mention red supergiants again in section 10.9.

The description of the evolution of the internal structures of high-mass stars is reasonably reliable. However, the description of their evolution in the H-R diagram (i.e., the way their observational appearance varies) on the assumption that such a star has a constant mass is misleading. Earlier we remarked that the most important single attribute of a star is its mass. Massive stars can lose so much mass through stellar winds, even on the main sequence, that their evolution is dramatically affected by this mass loss. To describe this evolution we must divide our massive stars into three subgroups.

The first subgroup consists of the most massive stars, those with masses greater than about 60 times that of the Sun. While in the hydrogen-burning phase they experience dramatic bouts of extreme mass loss. A spectacular example of such a star is the star eta Carina whose ejected envelope can be seen in the Hubble Space Telescope photograph shown in plate 2. (Such stars are commonly called luminous blue variables [LBVs] or Hubble-Sandage variables, but are often difficult to classify.) By the time helium burning has started, the mass loss is so great that nearly all the hydrogen-rich atmosphere has been removed and what is left is essentially a bare helium-burning core. As in the constant mass case, such a star moves redward in the H-R diagram for some time, but eventually the effects of mass loss reverse this movement and the star actually ends up blueward of its original main-sequence position. It is then called a Wolf-Rayet star. These massive stars never become red supergiants during their evolution.

The stars in our second subgroup have initial masses between 25 and 60 times that of the Sun. In these stars, the mass loss is not so dramatic as for the most massive stars, and they can evolve into the red supergiant regions of the H-R diagram. In the red supergiant phase, analogously to the red giant phase met earlier, dramatic mass loss occurs and the atmosphere of the star is removed. The bare helium-burning core then evolves blueward in the H-R diagram, ending up blueward of the original main-sequence position. It becomes a Wolf-Rayet star. The main difference between a Wolf-Rayet star formed by the evolution of a star initially having 25 to 60 $M_\odot$ and one formed by the evolution of a star with an initial mass greater than 60 $M_\odot$ is that the Wolf-Rayet star with the lower mass progenitor can be surrounded by material ejected during the red supergiant phase.

The final group of stars comprises those with masses less than about 25 $M_\odot$ and greater than about 8 $M_\odot$ each. These stars evolve redward into the red supergiant region, but the mass loss is never sufficient for them to move blueward in the H-R diagram. These stars remain as red supergiants that may or may not be variable.

Evolutionary tracks of three massive stars are shown in figure 3-6.

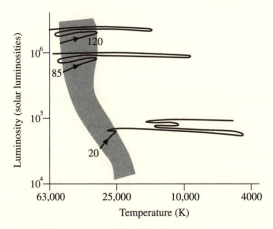

FIGURE 3-6   Evolutionary Paths of Massive Stars in the H-R Diagram. The paths of three stars with masses of 120, 85, and 20 times the mass of the Sun in the H-R diagram. The total evolution times range between millions of years for the 120 solar mass star and tens of millions of years for the 20 solar mass star.

The discussion of Type II supernovae in section 8.1 concerns the subsequent evolution of all massive stars.

SELECTED REFERENCES

Prialnik, D. *An Introduction to the Theory of Stellar Evolution*. Cambridge: Cambridge University Press, 2000.

Schönberner, D., and Blöcker, T. "Introduction to Stellar Evolution." In *The Molecular Astrophysics of Stars and Galaxies*, ed. T. W. Hartquist and D. A. Williams, 237–264. Oxford: Clarendon Press, 1998.

4

BASIC STRUCTURES OF WINDS AND WINDBLOWN BUBBLES

To understand winds, explosions, and the bubbles they blow, one must first possess a command of a number of simple concepts of the theory of hydrodynamics. This theory is used to describe flows in nonmagnetized liquids and gases in which ions, electrons, atoms, and molecules collide with each other on lengthscales that are short compared to lengthscales associated with structures in the flows. The flows that we observe as part of our ordinary experience (e.g. waves on lakes, brisk winds on early winter days, and flames on gas stoves) all obey the laws of hydrodynamics. The ways in which pressure differences affect flows, sound waves, and shocks are addressed in this chapter.

A steady flow is one that appears the same for a very long time. For instance, the flow of air around an airplane travelling at a constant speed and altitude on a calm day appears steady to the pilot. Steady wind solutions of the equations of hydrodynamics were first obtained by Eugene Parker. The nature of the flows described by these solutions are considered in this chapter before the basic notions of hydrodynamics are applied to the structures of bubbles blown by winds.

4.1 Thermal Pressure, Ram Pressure, Sound Waves, and Shocks

Thermal pressure is the force per unit area that a substance exerts on the surface surrounding it due to its having thermal energy. If other forces (including gravity and electric and magnetic forces) can be neglected, gas in a region of high thermal pressure will move toward regions of lower pressure. For instance, in an internal combustion engine, the burning of fuel creates a region with high thermal pressure, and the movement of gas in a direction allowing it to attain a lower thermal pressure drives the motion of the piston.

As noted in section 3.1, in an ordinary gas, the thermal pressure is proportional to the temperature and proportional to the number density. Temperature is proportional to the thermal energy per particle. The number density is just the number of particles in a unit volume (cf. section 2.5). Thus, the thermal pressure of a gas is proportional to its thermal energy density, which is the amount of thermal energy contained in a unit volume of the gas. For an atomic gas the thermal pressure is two-thirds the thermal energy density.

The thermal energy per atom in an atomic gas is one half an atom's mass times the square of the speed typifying the thermal motions (cf. section 2.6). This fact and the relationship between thermal pressure and thermal energy density give a relationship between the pressure and the typical thermal speed. The thermal pressure is roughly the product of the mass per particle, the number density of particles, and the square of the thermal speed.

The relationship between thermal pressure and the thermal speed suggests a way of defining a *pressure* associated with the large-scale motion of a gas. The *ram pressure* of a gas is the product of the mass per particle, the number density of particles, and the square of the speed of the large-scale motion. The ram pressure of a strong wind in one's face makes an uphill run more difficult. Figure 4-1 shows a region with high thermal pressure to the left separated by a thin but movable plate from a region of very low thermal pressure to the right. If the gas to the right of the plate is initially motionless and the plate is released, it will be driven to the right. However, if the gas to the right of the plate is driven (by a fan say) toward the plate, the plate's motion is retarded. Indeed, even if its thermal pressure is negligible, if the breeze or wind to the right is moving with a speed sufficient that its ram pressure is equal to the thermal pressure of the gas to the left of the plate, the plate will not move.

The famous Bernoulli's law, upon which the flight of airplanes is based, states that in a steady flow the sum of the thermal pressure and the ram pressure is constant. That such an important law can be expressed so succinctly once the concept of ram pressure is introduced indicates what a powerful concept it is.

A sound wave is a pressure disturbance that moves through the gas. The speed at which sound waves move relative to the gas is called the sound speed. A calculation of the sound speed shows that it differs from the typical ther-

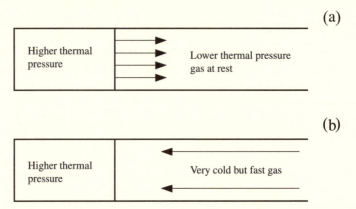

FIGURE 4-1   The Effects of Thermal Pressure and Ram Pressure. Panel (a) shows a cylinder open at its right end but with a fixed wall at its left end. The volume between the fixed wall and a thin plate contains gas with a thermal pressure that is higher than that of the gas to the right of the plate. Initially the plate was held fixed by pins attached to the cylinder and all gas was motionless. However, when the pins were removed, the higher thermal pressure of the region to the left of the plate caused the gas there to expand and induce the plate to move to the right. Panel (b) shows a cylinder in which very cold (i.e., very low thermal pressure) gas is injected at the open end with the speed required to make its ram pressure equal to the thermal pressure of the gas to the left of the plate. Even when the pins are removed, the ram pressure of the gas entering the cylinder from the right prevents the plate from moving.

mal speed, mentioned above, by at most 30 percent. The sound speed is a speed limit: the influence of pressure differences cannot spread more rapidly than the sound speed.

To understand some of the consequences of this speed limit, consider the sketches of flows around two rocket ships shown in figure 4-2. In each case, we will assume that we are piloting the ship so that from our perspective, it is at rest and the air is flowing toward and then around and away from the rocket. If a flow is at a speed less than the sound speed it is said to be *subsonic*. If the flow toward the rocket is subsonic, the disturbance in the flow caused by the presence of the rocket affects the flow far upstream; far upstream of the ship the flow starts to adjust to the rocket's existence. If a flow is at a speed greater than the sound speed it is said to be *supersonic*. If the flow toward the ship is supersonic, the information of the ship's location is not transmitted far upstream. If the flow is supersonic a shock develops near the ship and the upstream flow is unaffected until it reaches the shock.

(a)

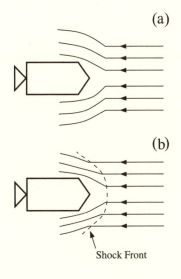

(b)

Shock Front

FIGURE 4-2   Subsonic and Supersonic Flows Around a Rocket Ship. (a) If the flow, as viewed from a rocket's cockpit, is everywhere slower than the speed of sound, the upstream flow can adjust smoothly as information about the rocket's shape can propagate upstream. (b) If the distant upstream flow appears to be supersonic, information about the rocket does not propagate upstream, and a shock exists around the rocket.

A shock is a thin region in which frictional effects in the gas convert much of the kinetic energy, associated with the motion of material flowing into it, to thermal energy. By thin, we mean that the thickness of the shock is small compared to any other lengthscales associated with the flow (e.g., in the case of a shock near a rocket, the shock is very thin compared to the length of the ship). As gas passes through the shock near the supersonic rocket it is slowed. The slowed gas is heated sufficiently that the sound speed in it increases above the speed at which it approaches the rocket.

Immediately behind a highly supersonic shock through which ionized gas (gas in which the electrons are not bound in neutral atoms) passes, the thermal energy per particle is about one half times the kinetic energy per upstream particle, as measured by someone looking upstream from the vantage point of the shock. This means that the immediate postshock temperature increases with the square of the speed of the upstream gas (cf. the fourth paragraph of section 2.6). Table 4-1 gives the temperature immediately behind shocks for different speeds of the upstream ionized gas. (To gain some appreciation of the magnitudes of these speeds, refer back to the penultimate paragraph of section 1.1; also recall that the speed of light is 300,000 km s^{-1}.)

The conversion of kinetic energy to heat in a shock causes the gas to slow down. As it slows, the gas gets denser. Ionized material passing at much more than the upstream sound speed through a shock will get denser by a factor of 4 and slow down by the same factor as it is heated.

TABLE 4-1   Temperatures of Shocked Ionized Hydrogen Gas

Shock Speed (km s^{-1})	Temperature (degrees Kelvin)
30	1.2×10^4
100	1.4×10^5
300	1.2×10^6
2,000	5.4×10^7
6,000	4.9×10^8
30,000	1.2×10^{10}

Figure 4-3 is a pictorial summary of some of the properties of a shock formed in ionized material moving at several times or more than the sound speed.

4.2 The Subsonic to Supersonic Transition in an Astronomical Wind

We assume the star or other astronomical source producing a wind to be spherical. We neglect all forces except those arising from spatial variations in thermal pressure and gravity.

A useful measure of the gravitational field strength is the escape speed, which is the minimum speed that material has to have to escape from a gravitating body. For instance, a spacecraft launched from the Earth's surface must have a speed of at least about 40,000 km per hour (a bit more than 10 km s^{-1}) in order to escape the Earth's gravitational field and travel to other planets. If the craft were launched from a space station about 25,000 km above the Earth's surface, its initial speed would need be only about 20,000 km per hour in order to escape the Earth's gravity.

As stated in the previous section, the thermal pressure is proportional to the number density times the temperature, and the temperature is just a measure of the thermal energy per particle. The temperature is proportional to the square of the sound speed. Thus, the sound speed is a measure of the pressure, just as the escape speed is a measure of the gravitational field strength.

At the surface of a star like the Sun the sound speed of the gas is lower than the escape speed. Even so, the pressure of the gas pushes upward and prevents the surface of the Sun from being sharply bound. Figure 4-4 shows schematically how the number density of gas falls near the surface of the Sun, due to the counteracting influences of the gas's pressure and gravity. When drawing figure 4-4, we assumed that the gas is static. Even if the gas at the surface is flowing outwardly, but at a speed less than the sound speed, the density profile near the surface is very similar to that shown in figure 4-4.

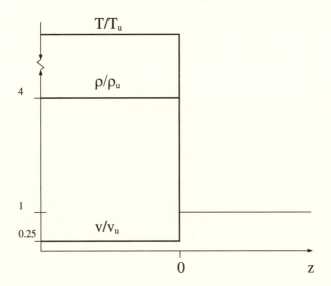

FIGURE 4-3 Shock Structure. We suppose that we are moving with the shock and see upstream gas moving toward us and gas that has passed through the shock move away from us. The shock is at $z = 0$. T, ρ, and v denote temperature, density (mass per volume), and speed, respectively. The subscript u signifies that the upstream value of the quantity is taken. Thus, upstream $T/T_u = \rho/\rho_u = v/v_u = 1$. We are considering a very fast flow in an ionized gas. The values of the increase in r and the decrease in v are shown explicitly. The large increase in T in the shock is indicated only schematically; the break in the vertical axis of the plot is included to imply that T/T_u behind the shock is very large compared to any number below the break.

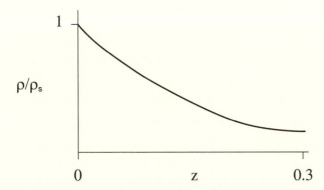

FIGURE 4-4 The Density of the Solar Corona. z is the distance above the Sun's surface measured in units of the Sun's radius, which is about 7×10^5 kilometers. ρ is the mass density of the hot ($T \approx 2 \times 10^6$ K) solar coronal gas above the cooler ($T \approx 6000$ K) surface of the Sun. ρ_s is the value of the corona's mass density at the surface of the Sun and is roughly 4×10^{-16} g cm^{-3}.

However, if there is a steady outflow (cf. section 1.2) from the surface of an object, at distances that are comparable to the size of the object or larger, the density profile of gas deviates substantially from that given by a simple extrapolation of the curve in figure 4-4. In fact, a study of the equations of hydrodynamics shows that a steady outflow that is subsonic near the surface of a spherical object must undergo a transition to being supersonic far away from the surface. The solutions demonstrate that the transition from subsonic to supersonic flow must occur at a particular spherical surface surrounding the object. At that surface the ratio of the escape speed to the sound speed has a particular value, which is fairly close to unity (but depends on the details of the thermal properties of the gas). Eugene Parker's early great contribution to the theory of steady astronomical winds was the discovery that the flow must undergo a subsonic to supersonic transition at the critical surface where the ratio of the escape speed to the sound speed has the critical value near unity.

Parker's finding can be justified in part with the following arguments based on physical intuition. The pressure of a gas acts to cause material to move from a region of high pressure to one of low pressure. Hence, the pressure of gas will cause material flowing subsonically outwardly from the Sun to accelerate. Once the acceleration leads to the outflow becoming supersonic, the pressure differences in the gas become less effective in causing acceleration (because the sound speed is a speed limit and the definition of supersonic flow is flow that is faster than the speed of sound). Thus, if the outflow were to become supersonic at a point where the escape speed is much greater than the sound speed, deceleration caused by the gravity would dominate and cause the gas to slow down again. If the subsonic to supersonic transition in the flow occurs where the sound speed is comparable to the escape speed, the gas manages to continue to flow (without decelerating) away.

Figure 4-5 shows the radial velocity (as a function of distance from a star with the same mass as the Sun) of a steady wind with a constant temperature of two million degrees Kelvin.

4.3 The Interaction of a Supersonic Wind with a Surrounding Medium

A steady astronomical wind undergoes a transition from moving subsonically to moving supersonically. As a wind moves away from its source, its speed becomes constant. When its speed is constant, the number density of a steady wind moving radially away from a source falls off as the inverse of the square of the distance to the source, just as the intensity of light drops in an inverse square law fashion.

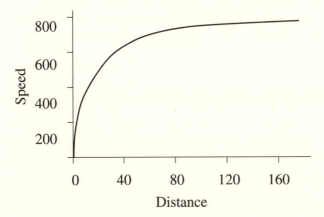

FIGURE 4-5   Expansion Speed of a Solar-Like Wind. The speed of a 2×10^6 K gas flowing in a wind away from a star like the Sun. The speed is given in km s^{-1}. The distance from the star, r, is in units 10^6 km. The Earth is at a distance of 1 AU (= 1.5×10^8 km) from the Sun.

As described in more detail in subsequent chapters, interstellar space is not a vacuum, and a wind will eventually encounter ambient material that will impede its flow. A bubble forms. As shown in figure 4-6, the bubble has four regions. From outside to inside, they contain (a) surrounding or ambient diffuse matter that is beyond the influence of the bubble, (b) ambient material that has been swept up by the bubble's expansion, (c) wind that has been decelerated by the interaction with the ambient matter, and (d) freely expanding supersonic wind that has not yet been decelerated. Many astronomical bubbles expand into the surrounding media at speeds much greater than the sound speeds in that ambient material; in such a bubble the surface separating regions (a) and (b) is the site of a shock moving into region (a). The surface between regions (b) and (c) is referred to as a contact discontinuity. An interesting question, remaining to be answered, concerns the degree to which swept-up material and slowed-down wind mix across a contact discontinuity. The surface between regions (c) and (d) is a shock through which the wind slows as it approaches the ambient material.

   If we consider the flow in a windblown bubble like that sketched in figure 4-6 from the perspective of someone moving with the contact disconti-nuity, we can apply Bernoulli's law (cf. paragraph 5 of section 4.1). Doing so, we conclude that if the far upstream ambient material is very dense com-pared to the wind material, the bubble expands into the ambient material at a speed that is about the speed of wind in region (d) times the square root of the ratio of the density at the outer edge of region (d) to the density in

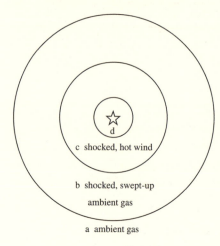

c shocked, hot wind

b shocked, swept-up
ambient gas

a ambient gas

FIGURE 4-6   The Structure of a Wind-Blown Bubble. Ambient material in region (a) is swept up by a shock at the outer edge of the expanding bubble. The surface separating region (a) from region (b), which contains swept-up ambient material, is the site of the shock. The wind is freely expanding in region (d) but passes through a shock that decelerates it at the boundary between region (d) and region (c), which contains hot shocked wind. The shocked ambient gas and shocked wind abut one another.

region (a). That is, the denser the ambient material is, the more slowly the bubble expands.

Since the speed of bubble expansion is generally much lower than the speed of the wind before it passes through the inner shock, the temperature of the shocked wind material is much higher than that of shocked ambient material (cf. table 4-1). Gas at temperatures above about 10 million degrees Kelvin does not lose energy very effectively by radiating it into the surrounding space, whereas gas at a temperature much below this does. Hence, in many situations, the material in region (c) of a bubble remains very hot, whereas material in region (b) cools (often to about 10,000 degrees Kelvin). The cooling in region (b) and the fact that the thermal pressure is fairly constant across regions (b) and (c) result in material in region (b) being compressed to a density much greater than that in region (a); hence, region (b) is a relatively cool, thin shell. Region (c) contains hot, tenuous gas.

4.4 Back Briefly to the Solar Wind

Parker's work on the structure and dynamics of winds was motivated by an interest in the solar wind. Although many of the concepts addressed in this chapter are of relevance to many types of winds, we will focus here on the solar wind.

Direct measurements of its properties were made for the first time in the mid-1960s when spacecraft could be launched far enough from the Earth. The properties of the Sun's wind vary with the solar cycle (cf. section 1.1's mention of the 11-year cycle of sunspot number) but it is possible to quote typical values of the parameters that characterize the wind. At the distance of the Earth from the Sun (which is 1 AU; cf. section 2.3), the solar wind carries about seven protons and seven electrons in every cubic centimeter at a speed of about 450 km s^{-1}. Typically the temperature is about 100,000 degrees Kelvin, very roughly about 10 times lower than the temperature of the wind in the vicinity of the Sun.

The Earth is located within region (d) (cf. figure 4-6) of the solar wind-blown bubble. That is, the solar wind has not passed through and been slowed down by a shock (cf. the boundary between regions (c) and (d) in figure 4-6). From estimates of the properties of the local interstellar medium and from the measured properties of the solar wind at Earth, we can infer that the shock that decelerates the solar wind is at a distance roughly 100 times greater from the Sun than the Earth is. This value of its distance from the Sun is somewhat uncertain and depends on the theoretical model that is adopted in the calculation. In any case, the solar wind lies between us and what might be considered true interstellar space.

SELECTED REFERENCES

Dyson, J. E., and Williams, D. A. *The Physics of the Interstellar Medium.* 2nd ed. Bristol: Institute of Physics Publishing, 1997.

Hundhausen, A. J. "The Solar Wind." In *Introduction to Space Physics*, ed. M. G. Kivelson and C. T. Russell, 91–128. Cambridge: Cambridge University Press, 1995.

5

STAR FORMATION AND LOW-MASS
YOUNG STELLAR OBJECTS

Rotation, in conjunction with a magnetic field, drives the outflow of a low-mass young stellar object (LMYSO), a young star with a mass a few times that of the Sun or less. The star formation process determines how a LMYSO rotates. Consequently, the nature of the outflow is intimately connected with the entire history of the LMYSO's formation. Therefore, in this chapter we begin with a review of observations of giant molecular clouds (GMCs), which are the sites of most stellar birth in the Galaxy, and also provide an overview of the standard picture of the formation of a LMYSO.

Magnetic fields play a central role in controlling the rate of collapse in LMYSO formation, as well as in the collimation or channelling of the outflow from one. (The collimation can give rise to a jet!) Thus, this chapter contains an introduction to the properties of magnetic fields.

After treating the mechanism inducing mass loss from a LMYSO and the effects of such mass loss on surrounding material, we shall turn in the next chapter to high-mass young stellar objects (HMYSOs). The births of high-mass stars differ considerably from those of low-mass stars. The mechanism by which a HMYSO loses mass may well also differ substantially from the process by which a LMYSO drives outflow. In any case, the outflows of the largest young stars affect the structures of the GMCs in which they form on scales that are a good fraction of the size of a GMC and induce additional star formation.

5.1 The Interstellar Gas and Giant Molecular Clouds

Roughly 1 billion solar masses, or 1 percent of the matter in the Galaxy, is contained in a diffuse interstellar medium with an average number density of one hydrogen nucleus per cubic centimeter. Much of the neutral atomic gas is at temperatures around 100 K and in clouds having masses roughly 100 to 1,000 times that of the Sun. Typically these clouds contain about 30 atoms per cubic centimeter. In the Milky Way the 100 K clouds fill a few percent of the volume of interstellar space.

Such clouds are not massive enough for their own gravitational fields to keep them bound (cf. the description of the Jeans mass in section 5.3). Rather, hotter, more tenuous material surrounds the clouds and confines them with its pressure. The hotter material is usually in either of two states. In one, the gas is at about 10,000 K and has a number density of a few hydrogen nuclei in every 10 cubic centimeters. In the other state, the gas is at temperatures of around 1 million degrees Kelvin and has a number density of about a hydrogen nucleus in every thousand cubic centimeters. The 10,000 K gas is heated primarily by the ultraviolet radiation of stars and soft X-rays emitted by the 1 million K gas, whereas the hottest gas is energized by the explosions (called supernovae; cf. section 8.1) of massive evolved stars. The ultraviolet radiation has wavelengths around 100 nm; the soft X-rays have wavelengths of about 1 to 10 nm.

The pressure of all of the diffuse gas mentioned so far is usually around 10^{-17} times that of the Earth's atmosphere at sea level. However, this low pressure gas is detectable in various ways. The atomic hydrogen in the clouds emits radiation with a wavelength of 21 cm. The 21 cm radio radiation has been mapped in the Galaxy and in other galaxies.

Figure 5-1 shows an optical image of a nearby galaxy. It, like the Milky Way, is a spiral galaxy, and in figure 5-1 the spiral nature of the galaxy is clear. The theory of the origin of spiral structure in some galaxies is incomplete. The spiral arms may be regions with only modest density enhancements and may show up so prominently because only a slight increase of the density above that in the interarm regions is required to induce much higher rates of star formation. The arms are thought by some researchers to be waves of somewhat enhanced density propagating around a galaxy at a rate differing slightly from the rotation rate.

Due to a spiral density wave's density being higher than that typical of most of a galaxy, such a wave has an enhanced gravitational field. The gravitational field of a passing spiral density wave may lead to the formation of GMCs by causing a large number (around 100 to 1,000) of atomic clouds to fall toward one another.

One of the most studied GMCs is the Rosette Molecular Cloud (RMC). Figure 5-2 shows a map of CO emission from it. (The emission arises in tran-

FIGURE 5-1   An Optical Image of a Dusty Spiral Galaxy (NGC 4414). Image courtesy of NASA/STScI.

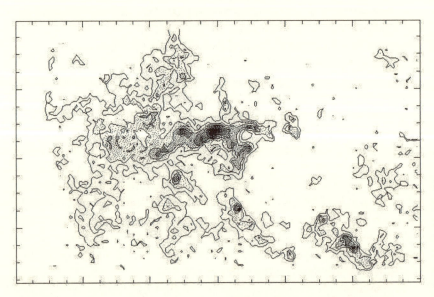

FIGURE 5-2   Map of the CO Emission from the Rosette Molecular Cloud. The total extent of the map is about 300 light years by 200 light years. The brightest emission comes from the regions surrounded by the most contours. The clumpiness of the cloud is apparent. J. P. Williams kindly provided the figure. The observations are described in detail in the paper by Williams, J. P., Blitz, L., and Stark, A. A., "The Density Structure of the Rosette Molecular Cloud: Signposts of Evolution," *Astrophysical Journal* 451 (1995): 252–274.

sitions resulting in the molecule rotating more slowly; collisions between molecules create the more rapidly rotating molecules.) The analysis of such maps shows that the RMC contains about 70 clumps ranging in mass from tens of solar masses to thousands of solar masses. In all, the RMC contains about 100,000 $M_\odot$ of material and has a maximum linear extent of about 200 to 300 light years. The average number density in a clump is roughly 200 hydrogen molecules per cubic centimeter, but densities several times higher are also common. The gas temperature is around 10 K. Most of these clumps are "translucent" in the sense that only a few to about 30 percent of any visual radiation incident upon them can pass through. The translucent clumps constitute a good starting point for a treatment of star formation (cf. section 5.3). At present, the nature of the material between the clumps remains uncertain.

The Galaxy contains thousands of GMCs, which altogether contain around 1 billion solar masses of matter. The GMCs are major sites of stellar birth. The RMC has eight clumps in which young stars are known to be embedded. Observations of infrared radiation with wavelengths of tens to hundreds of microns led to the discovery of those young stars. All of the clumps contain dust, and if a clump contains a star, the dust is heated by the star's radiation and emits in the infrared. The dust grains also prevent ultraviolet and optical radiation due to stars outside the clumps from penetrating to the centers of the thickest clumps. All eight of the clumps in the RMC with embedded stars are amongst the sixteen thickest or darkest of the RMC's clumps.

Motions within a clump are measured from the way that they redshift and blueshift radiation emitted by the molecules in it. A transition between any type of bound states occurs at a fixed wavelength, as a consequence of the quantum laws of Nature. For example, the transition between the two lowest rotational states of CO has a distinct wavelength of 2.6 mm. As described in section 1.3, motion toward the observer leads to a perceived decrease in the wavelength of the radiation, whereas motion away from the observer causes a perceived increase in the wavelength. If the gas in a cloud is moving to and fro in a random way about the cloud's average position, some of the radiation from the cloud will be blueshifted and some redshifted, causing the broadening of the wavelength range of the received radiation. From such broadening of emission from clumps, we infer that the gas in an RMC-type clump moves randomly at a speed of a couple kilometers per second relative to the clump's average speed. One way to cause gas to have a random component to its velocity is to heat it; random motions associated with the average thermal energy per particle in a cloud (which is measured by the temperature; cf. sections 2.5 and 4.1) give rise to broadening. However, the observed broadening of CO emission from the clumps is several tens of times that expected to be due to thermal broadening, an issue considered in section 5.3.

5.2 The Basic Physics of Magnetic Fields and Magnetized Fluids

The properties of atoms and molecules are affected if they are placed in a magnetic field. Shifts of the energies of some of the bound states of the atoms and molecules will occur, causing the splitting of some spectral features into multiplets—that is, the presence of a magnetic field can cause the emission from an atom or a molecule that normally occurs at one particular wavelength to occur at two or more slightly different wavelengths. Observations of such splitting in some features allows the measurement of magnetic field strengths in astronomical sources. Splitting in features originating in interstellar clouds implies that they contain magnetic fields of sufficient strength to influence the structures and dynamics of the clouds. Magnetic fields play major roles in star formation and in the winds of LMYSOs. Hence, we consider the basic physics of magnetic fields in this section.

Imagine that a number of identical aligned magnets are brought into the same volume. The total force on all magnets is proportional to the product of the number of magnets in the volume with the total local magnetic field strength. The total local magnetic field strength is proportional to the number of magnets in the volume. Thus, the total magnetic force on all of the magnets in the volume varies as the square of the local magnetic field strength.

However, the magnetic force per unit volume depends on more than the local magnetic field strength alone. It also depends on how the magnetic field varies with position. To see this, consider a magnet aligned with and equidistant from two other magnets identical in all ways to it. The middle magnet experiences equal and opposite forces due to the other magnets, so that there is no net force on the middle magnet. However, if one of the other magnets is replaced by two identical magnets, the middle magnet will experience a force pushing it in the direction with fewer magnets. The magnetic force acts in the direction of lowest magnetic field strength. Note how similar the behaviour of this magnetic force is to the force exerted by thermal pressure (cf. section 2.1). In fact, it is usual to define the *magnetic pressure*, which is proportional to the square of the local magnetic field strength. If the magnetic field is everywhere in the same direction, the net magnetic force exerted on a volume of material is calculated the same way from the magnetic pressure as the thermal pressure's force per unit volume is calculated from the thermal pressure.

Figure 5-3 gives an indication of how the magnetic pressure generates a force. If the magnetic field is uniform there is no net magnetic force on the magnetized medium. The concept of a magnetic field line is used in figure 5-3. The strength of the field is proportional to the number of field lines passing through a region, and the direction of the field is parallel to the field lines.

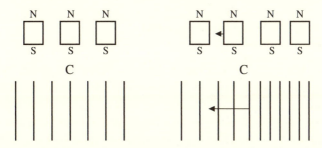

FIGURE 5-3   The Effect of Magnetic Pressure. The magnetic field direction is indicated by the direction of the lines and the strength is proportional to the number of lines passing through an area. In the left-hand side of the figure, the magnetic field is uniform and there is no net force arising from the magnetic pressure since it, too, is uniform. Magnets distributed at equal intervals are depicted above these lines; in such a situation, there is no net force on the middle magnet. In the right-hand side of the figure, the magnetic field is not uniform but is stronger to the right. Consequently, the magnetic pressure increases to the right, resulting in a magnetic force to the left (i.e., in the direction of decreasing magnetic pressure, exactly analogous to the way gas thermal pressure acts in the direction of decreasing gas pressure). Four magnets are shown above these field lines. The two to the right are closer together, giving rise to a stronger magnetic field to the right than to the left, resulting in a magnetic force to the left.

The spatial variation of the magnetic pressure contributes only one of the two types of magnetic forces that can act on a magnetized medium. The other type of magnetic force on a material is associated with the *magnetic tension*. To understand it, consider the situation pictured in figure 5-4. One of its panels depicts a plate of highly conductive matter that has been at rest between a large magnet and an identical large magnet just above it. (A *highly conductive* or *perfectly conductive* material is one in which electric charges move very readily in response to an electric field.) The magnetic field goes directly from the North Pole of the lower magnet to the South Pole of the upper magnet and is constant through the plate. A forceful push of the plate is required to move it perpendicular to the field. The field is stretched in the direction that the plate has moved. Just as a rubber rope would exert a force to restore itself to a straight shape, the magnetic field is tense and exerts a restoring force. Bending or twisting a magnetic field causes it to produce a tension force per unit volume. The *tension force* per unit volume is proportional to the product of the magnetic field component in the direction of the twisting or stretching with the magnetic field component in the direction perpendicular to the direction of the twisting or stretching. The tension force per unit volume also depends on the lengthscale over which the bending and twisting takes place, so that a field

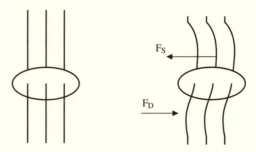

FIGURE 5-4   The Magnetic Tension Force. To the left is shown a conducting plate through which a uniform magnetic field passes. If the plate were subjected to a force perpendicular to the field lines, the plate would move and distort the field lines by bending them. F_D in the picture to the right represents the force causing the distortion of the magnetic field. The bending or stretching of the field lines produces a magnetic tension force indicated by F_S.

with a sharp bend or twist in it is more stressed than a gently bent or twisted field; the tension force per unit volume is inversely proportional to the lengthscale of the distortion.

As is clear from the above paragraph, motions of conducting material threaded with a magnetic field change the field. If the motion is perpendicular to the field and causes a compression, the field strength increases in proportion to the density. Motion along the magnetic field does not change the magnetic field, but differential motion (often called *shear*) perpendicular to the magnetic field causes it to bend or twist.

In section 4.1 we described sound waves. The presence of a magnetic field changes the nature of the waves that exist and, thus, changes the limiting speed at which information can be transmitted without the development of a shock. The *Alfvén speed* is the speed at which one type of wave propagates in a magnetized medium. The magnetic pressure is given by one half the mass density times the square of the Alfvén speed, just as the thermal pressure is given by the product of the mass density and the square of the sound speed multiplied by a factor of order unity (cf. sections 4.1 and 4.2).

Figure 5-5 illustrates the nature of an Alfvén wave (named after the Swedish scientist who predicted its existence) propagating in the direction of the large-scale magnetic field. It propagates at the Alfvén speed. Its driving force is magnetic field tension, just as tension on a whip caused by a rapid movement of its grip is the force that induces a disturbance to propagate to the whip's tip. In an Alfvén wave, no compression of gas occurs; instead, all material in any given plane perpendicular to the wave propagation rotates at the same rate around the propagation direction. As no compression occurs, the thermal pressure plays no role in the propagation of an Alfvén wave.

In contrast, as described in section 2.1, sound waves are associated with disturbances in the thermal pressure and compression in the gas. The nature of sound waves is modified by the presence of a magnetic field, and to mark this modification sound waves in magnetized media are called *magnetosonic waves*. Consider a magnetosonic wave propagating perpendicular to the large-scale magnetic field. Where it causes the compression of gas, the thermal pressure *and* the magnetic pressure increase. The actions of the two types of pressure together cause the response of the gas to a disturbance to be faster than if only one type of pressure were important—hence, what is called the *fast-mode magnetosonic speed* is higher than either the ordinary sound speed in a nonmagnetized medium or the Alfvén speed. In fact, for waves propagating perpendicular to the magnetic field, the fast-mode speed is the square root of the sum of the squares of the Alfvén speed and the ordinary sound speed. The fast-mode speed is the fastest speed at which infor-

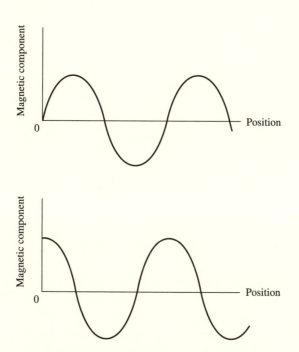

FIGURE 5-5   The Magnetic Field in an Alfvén Wave Propagating Along the Direction of a Uniform Background Magnetic Field. The magnetic field components along two directions perpendicular to each other are shown. Both of these directions are perpendicular to the propagation direction. When one component has a maximum value, the other component has a value of zero. The disturbance of the magnetic field associated with the wave rotates around the propagation direction.

mation can be transmitted without the development of a shock. If the magnetic pressure is much less than the thermal pressure, the information speed limit differs little from the ordinary sound speed. However, if the magnetic pressure is much greater than the thermal pressure, that speed limit is much larger than the ordinary sound speed and marginally greater than the Alfvén speed.

5.3 Rotation and Support of Translucent Clumps in Giant Molecular Clouds: The Roles of Magnetic Fields

As stated at the start of the chapter, rotation is important for mass loss from LMYSOs, and the rotational properties of a LMYSO are determined by the history of its formation. One of the major ways that magnetic fields affect star formation is the manner in which they transfer angular momentum.

Angular momentum is a measure of the product of the mass of an object, the velocity at which it rotates, and the typical distance of its material from the axis about which the object rotates. Unless some force acts to tie the rotating object to other material, its angular momentum will be conserved. Angular momentum conservation is responsible for the increase in a spinning ice skater's rotation rate as the skater's arms are drawn closer to the body.

Our Galaxy rotates, and the rotation rate depends on the distance from the Galactic Center. This distance dependence of the Galactic rotation rate results in the rotation of any very extended object about an axis through its center-of-mass. Material collapsing to become a GMC makes up such an extended object and may be referred to as a proto-GMC. If angular momentum were not somehow transported away from a proto-GMC, the object would spin faster and faster as it collapsed. The galactic magnetic field passes through a proto-GMC, and carries angular momentum away from it during its collapse. To see how this occurs, refer to figure 5-6, which depicts a rotating object and its surroundings through which a magnetic field passes. The rotation causes a bending of magnetic field lines. This results in magnetic tension, which resists the increase in the rotation rate of matter near the rotation axis while causing material further from the axis to rotate faster. As collapse occurs the object loses angular momentum. A GMC rotates about an axis through it on a timescale of about the rotation period of the Galaxy; this timescale is comparable to, or even longer than, the expected lifetime of a GMC, and a GMC does not complete more than a fraction of a rotation about itself before its life is over.

The magnetic field couples the rotation of each individual translucent clump in a GMC to the GMC as a whole. Thus, the rotation rate of each translucent clump is low.

Magnetic tension and variations in the magnetic pressure in a translucent clump contribute to the support of it against its own gravity. A review

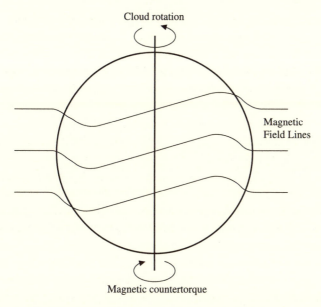

FIGURE 5-6   The Distortions of Magnetic Field by Cloud Rotation. A tension force is generated, resulting in the production of a magnetic torque decelerating the cloud's rotation.

of the issue of the stability of nonmagnetized clumps is necessary before a treatment of magnetic support of clumps.

A nonmagnetized and nonrotating clump will undergo gravitationally driven collapse if its gravitational escape speed has a value that is about twice the sound speed in the clump or higher. This makes sense because if the sound speed is high, the speed associated with random thermal motions is high and causes expansion. The gravitational escape speed for a cloud with a specified mass is higher the smaller the cloud is in spatial extent, or equivalently the denser the cloud is. Thus, the escape speed can be expressed in terms of the cloud mass and density. Consequently, we are able to calculate from its density and its sound speed, the minimum mass of a cloud (with a given temperature and number density of particles) that will collapse due to its own gravity. As the sound speed depends on temperature, we may use either the temperature or the sound speed in the calculation. The minimum mass for collapse is called the Jeans mass and increases with temperature and decreases with density. If the Jeans mass is expressed in terms of the sound speed and pressure, one finds that it increases with the sound speed to the fourth power and decreases as the inverse of the square root of the pressure. The pressure on a low mass clump must be high in order for gravity to cause it to collapse.

In the translucent clumps in giant molecular clouds, the temperature is about 10 K, corresponding to a sound speed of about 0.2 km s^{-1}. For a number density of hydrogen molecules of roughly several hundred per cubic centimeter, the Jeans mass is about 10 $M_\odot$. However, we have neglected magnetic forces!

If the magnetic field pressure is considerably greater than the thermal pressure, the minimum mass of a cloud that will collapse due to its own gravity may be estimated from the equation for the Jeans mass through the substitution of the Alfvén speed and magnetic pressure for the sound speed and ordinary (thermal) pressure, respectively. In translucent clouds the Alfvén speed is about 10 times the sound speed, and the magnetic pressure is about 100 times the thermal pressure, giving a Jeans mass closer to 10,000 $M_\odot$ than to 10 $M_\odot$. A few of the RMC clumps have masses of over 1,000 $M_\odot$. Given the uncertainty in observationally derived numbers these few largest clumps may be able, once triggered, to undergo fairly direct gravitationally driven collapse to form stars, but the less massive ones will not.

That the less massive ones will not may seem puzzling. After all, if the pressures of the surroundings increase, the densities of the clumps increase. For a nonmagnetized clump, as long as the sound speed stays constant, a sufficient increase in the pressure will cause its own gravity to drive collapse to stellar densities. Shouldn't this be the case with magnetized clumps as well? The answer to this question is "No." Increasing the external pressure on a clump and causing it to go to higher density also causes the magnetic field to get stronger. Though the cloud is denser and its gravity stronger, the increasing magnetic pressure in the clump will prevent gravity from winning unless either the gravity was winning from the very beginning or the clump somehow sheds some magnetic field.

A cold magnetized cloud with a mass below the minimum one required for gravity to drive its collapse is said to be *magnetically subcritical*. Magnetically subcritical objects become the birthplaces of LMYSOs. We shall describe later how such an object can shed some magnetic field.

A cold, magnetized clump with a mass greater than the minimum mass required for it to collapse in response to its own gravity is said to be *magnetically supercritical*. Initially, such a clump may be warm enough (or as discussed below, contain enough waves) that its gravity will not cause it to collapse. However, a progressive increase in the external pressure on it would cause it to collapse. Indeed, pressure increases around magnetically supercritical clumps are thought to trigger the births of HMYSOs.

So far we have considered the support of clumps by thermal pressure and by forces associated with the large-scale magnetic field. By *large-scale* magnetic field, we mean that part of the magnetic field that varies only on lengthscales comparable to or larger than the length of the clump or cloud. Yet another source of support is also important in translucent GMC clumps. It is associated with the random motions responsible for the broadening of the emission features. These random motions occur on scales much larger than the lengthscale over which a molecule typically moves before colliding with another molecule and, hence, are not thermal motions. They occur on scales that are also smaller than the lengthscale associated with a clump as a whole. They are random motions on intermediate scales, and the word *turbulence* is often associated with them. The turbulence in the translucent GMC clumps consists of a superposition of waves. The speed associated with the broadening of a line by redshifting and blueshifting due to the motions is a lower bound to the Alfvén speed in a cloud. Waves giving rise to apparent random motions faster than the Alfvén speed damp away quickly and are not present.

Anyone who has tried to stand in knee-deep water as a wave has passed knows that the wave exerts a force as it breaks against one's legs. The waves in a clump exert forces that help support the clump. A cold nonrotating clump with a strong magnetic field but no internal turbulence will be flattened along the magnetic field direction. A cold nonrotating clump with a strong magnetic field and turbulence with an associated speed comparable to the Alfvén speed will not be greatly flattened because the turbulence helps puff it up along the field direction.

5.4 The Collapse of a Translucent Giant Molecular Cloud Clump

In the previous section, we described how pressure and tension forces caused by the average large-scale magnetic field contribute to the support of clumps, and also the role that smaller-scale waves or fluctuations in the magnetic field play in support. If the magnetic fluctuations are comparable in strength to the large-scale magnetic field, they prevent the clump from being very flattened in a direction parallel to the typical direction of the large-scale magnetic field. However, if the magnetic fluctuations in a clump with a much higher magnetic pressure than thermal pressure were damped, it would become flattened.

One means of damping the fluctuations is through ion-neutral friction. The translucent clumps in a GMC are primarily neutral, and (as described below) the fraction of particles that are charged is somewhere between one millionth to one ten thousandth, depending on the depth into the clump. Magnetic forces act directly on charged particles, but neutral atoms and molecules experience no net magnetic force. This means that in a fluctuating magnetic field, such as that associated with an Alfvén wave, the charged particles respond directly to the magnetic force, and move relative to the neutrals. Collisions between the charged particles and neutrals give rise to friction, and the frictional coupling causes the neutrals to move as well, but at a velocity not quite equal to the velocity of the charged particles. Figure 5-7 illustrates the relative motion. Most of the friction is due to collisions between ions and neutrals, whereas the friction arising due to electron-neutral collisions is negligible; hence, we refer to ion-neutral friction and ion-neutral coupling, even though electric forces cause ions and electrons to keep together. The difference between the ion and neutral velocities gets larger if the number of ions drops, as fewer particles imply fewer collisions. This increase in the relative velocity causes an increase in the rate at which the energy of the wave is converted into heat (because the energy transfer rate goes up more quickly with relative velocity than the frictional force does). Thus, a wave in a magnetized medium will damp faster due to ion-neutral friction if the fractional ionization (fraction of particles that are ions) is low than if it is high.

Ion-neutral frictional damping of waves in translucent clumps in GMCs may be responsible for their eventual flattening and gives rise to a natural maximum thickness for such clumps. The possibility of a maximum thickness follows from the fact that the fractional ionization drops from the edge of a clump to its center due to the exclusion (by dust) of ultraviolet radiation capable of ionizing material. Figure 5-8 shows the fractional ionization,

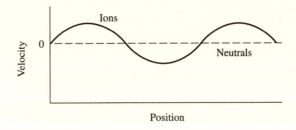

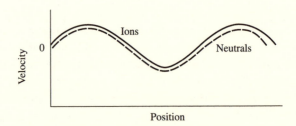

FIGURE 5-7 The Relative Motions of Ions and Neutrals in a Magnetohydrodynamic Wave in a Translucent Clump. In the upper panel, the motion of the ions is depicted under the assumption that there are no collisions between ions and neutrals. The lower panel shows a more realistic situation in which collisions result in the neutrals moving at a velocity differing somewhat from that of the ions.

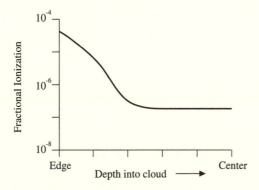

FIGURE 5-8   The Fraction of Particles That Are Ions as a Function of Depth into a Cloud (or Clump).

calculated with a theoretical model of the chemistry, in a region with a density around that of some of the denser RMC translucent clumps. Notice the region of rapid decrease in fractional ionization with increasing depth. Clumps that are not too thick will have a fractional ionization that varies gradually with position. Thicker clumps will have a much lower fractional ionization at their centers than at their edges. Since waves damp more rapidly due to ion-neutral friction in regions of low fractional ionization, clumps that are thicker are not turbulent at their centers. The lack of significant turbulence and wave activity at their centers implies that thick clumps lack support due to the waves and begin to collapse and flatten among the directions of their large-scale magnetic fields.

As stated above, the exclusion of radiation from the center of a clump is what causes a drop in fractional ionization. Consequently clumps with dark centers are those with low central wave activity. A number of processes may be responsible for initiating the darkening of a clump and the subsequent decrease in support due to waves and collapse. For instance, the wind of a star may turn on and blow a bubble in a GMC; the associated increase in pressure then causes any clump to be compressed, and this compression causes the center of the clump to be darker than it was before.

It is likely that a magnetically supercritical clump continues to collapse due to its own gravity once wave support in it ceases to be important. The collapse of a magnetically supercritical clump is thought to lead directly to the formation of an HMYSO. Once its wind turns on, the new HMYSO may induce the darkening, subsequent wave damping in, and collapse of, more clumps. If plenty of magnetically supercritical clumps are around, then the formation of HMYSOs may be a self-propagating phenomena.

An increase of the pressure on the outside of a magnetically subcritical clump will cause it to darken, and if it darkens sufficiently, damping of waves in it will allow some additional compression. However, the additional compression will not immediately lead to the gravity of the clump dominating and inducing collapse to form a star. Objects known as low-mass dense cores are probably formed as a consequence of the compression of magnetically subcritical clumps.

5.5 Low-Mass Dense Cores and Their Collapse

Dense cores are considered to be the immediate precursors of stars. Those involved in the formation of low-mass stars are probably fragments of initially translucent, magnetically subcritical clumps. Each low-mass dense core is supported primarily by its large-scale magnetic field and contains no significant wave activity. Typically, one has a mass of a few to a couple of tens of solar masses, a temperature of 10 to 30 K, a number density of around 10,000 to 100,000 hydrogen molecules per cubic centimeter throughout most of its volume, and a lengthscale of around a few tenths of a light year. Figure 5-9 shows the distribution of low-mass dense cores in the Taurus-Auriga

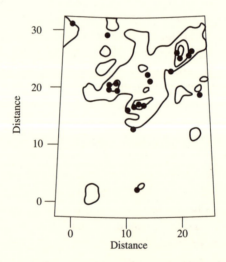

FIGURE 5-9   The Distribution of CO and Dense Cores in Part of the Taurus-Auriga Region. High concentrations of CO emission are indicated by contours. The locations of dense cores bright in other emissions, including NH_3, are indicated by dots. The distance is measured in parsecs (one pc is about three light years). Figure adapted from Myers, P. C. ,"Dense Cores and Young Stars in Dark Clouds," in IAU Symposium No. 115, *Star Forming Regions*, ed. M. Peimbert and J. Jugaku, 33–43 (Dordrecht: D. Reidel Publishing Company, 1986).

complex, which (at a distance of about 500 light years) is the major region of star formation closest to the Sun.

The magnetic coupling of a dense core to its surroundings restricts its rotation rate to roughly that of the Galaxy (cf. the third through fifth paragraphs of section 5.3). However, the magnetic field does not prevent collapse due to gravity or the spin-up of collapsing material forever. Rather, much of the magnetic field slips through a dense core as illustrated in figure 5-10. As mentioned in the previous section, the magnetic force acts directly on charged particles but not on neutral particles, causing the ions and electrons to move relative to the neutrals. The magnetic force drives the charged particles outward, and the magnetic field weakens as the charged particles move outward. This allows a *slow collapse* of the neutrals as the magnetic field weakens. The weakening of the magnetic field and the collapse of the neutrals takes place on a timescale that is proportional to the fractional ionization. For typical low-mass dense cores that timescale is about a million years, which is a few times longer than a core would require to collapse if magnetic forces were negligible.

As collapse of a dense core advances, the fractional ionization drops even further. Eventually the fractional ionization becomes so low that friction between ions and neutrals has almost no effect on the dynamics of the neutrals, and the presence of a magnetic field can be completely neglected in a consideration of their dynamics. At this time, the angular momentum of the neutral material is no longer affected by the magnetic field, and further collapse leads to a spin-up of the neutral material. The spin-up from a rotation rate similar to that of the Galaxy (i.e., once roughly every several hundred million years) begins to occur at densities only somewhat in excess of those of the densest of the low-mass dense cores. The spin-up prevents the material from falling all the way to the axis about which the rotation takes place, though collapse parallel to the rotation axis is not restricted by the spin. As a consequence, infalling material accretes in a disk with a radius of order 100 AU (see section 2.2 for the definition of an AU, or astronomical unit).

The inner boundary of a disk moves inward and the outer boundary moves outward as a consequence of friction in the disk. This friction is due purely to interactions of neutral particles with other neutral particles and, thus, differs from the ion-neutral friction discussed earlier. The friction in the disk arises because material at smaller distances from the rotation axis travels faster than material at greater distances, just as the Earth orbits around the Sun faster than the more distant Jupiter does. The friction slows the faster material and accelerates the slower material, causing the faster material to fall inward and the slower material to move outward. The inwardly moving material falls into a central region to form a protostar.

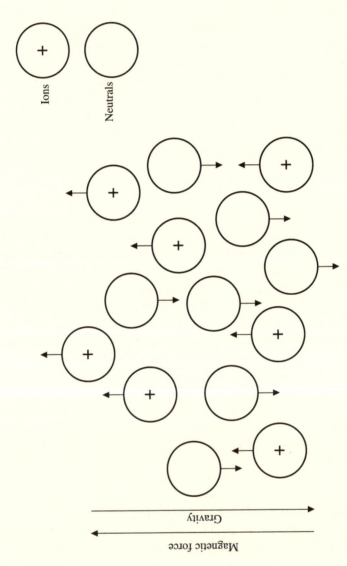

FIGURE 5-10 The Slip of a Magnetic Field out of a Dense Core. The magnetic force drives the relative motion of ions outward as gravity pulls the neutrals inward. The magnetic field is carried by the ions and other charged particles and weakens in the core as they move out.

5.6 Mass-Loss from a Low-Mass Young Stellar Object and Its Disk

In the inner region of a disk the friction leads to significant heating. At a distance from the central protostar not much less than the distance of Mercury from the Sun, the temperature in the disk around a protostar, with a mass comparable to that of the Sun, is high enough to melt most material. It is even high enough for collisions to ionize several elements, such as potassium and sodium, almost completely. The ionization in these regions, which produces charged particles, allows magnetic fields to become important again.

In the Sun the magnetic field is generated as a consequence of different rates of rotation at different distances from the Sun's center and turbulent convective motions below the solar surface. A similar process generates a magnetic field in a protostar and some regions in its disk.

The combination of rotation and a magnetic field results in a wind from the region near the inner boundary of the accretion disk. At present, a detailed theoretical understanding of the time-dependent development of the magnetic field and flow structure around a low-mass protostar and its disk does not exist. However, it is easier to understand the magnetic field and flow structures that the system finally attains. We now give a description of a somewhat fictional evolutionary picture that would result in final magnetic field and flow structures like those that are favoured for a low-mass protostar-disk system by some theoretical astrophysicists. We use this picture because we hope that the consideration of it helps in the understanding of the nature of the final structures, but we stress that the evolutionary picture is likely to be less realistic than the description of the final structures. The fictional evolutionary scenario is depicted in figure 5-11.

First consider a star with a well-developed large-scale magnetic field that resembles that of the Sun. Suppose that the inner edge of a disk is moving toward it due to the friction in the disk. Also suppose that the disk is a perfect conductor. Then as the disk's inner boundary moves closer to the star, it will push the magnetic field near it inward, as indicated in the upper part of figure 5-11. (This inward distortion of the field arises for the same reasons that the magnetic field is bent when a conducting plate between two magnets is moved as depicted in figure 5-4.) The field lines respond in such a way that they extend outwardly from the inner edge of the disk and at large distances follow roughly the same paths they would if the disk were not present.

A real disk is not fully ionized, nor is it a perfect conductor, and as a result, the magnetic field from the star can slip partially into the disk as it approaches the star. Thus, near the inner boundary of the disk, there is a region where magnetic field lines that extend far away from the star penetrate the disk. The lower half of figure 5-11 shows the magnetic field configuration that results as the disk moves further in. Some of the field lines are stretched to very great distances because some of the inner disk material is blown outward as a consequence of the magnetic field's tendency to rotate at the rate of the star. That tendency arises because though some field lines do get bent far away from the star, they turn back to the star and enter it near its poles. The stellar rotation rate is higher than that throughout most of the disk, and the interaction of the magnetic field with the material in the disk will accelerate it, causing it to move outward. This initiates the wind.

In a real system the disk does not approach a preexisting star. Rather, the star forms from material at the center of the disk, and the inner edge of the disk moves outward during the initial phases of protostellar development. Nonetheless, it is believed by many theorists that this sort of evolution leads to magnetic and flow structures similar to those attained in the evolutionary scenario described immediately above. Solutions to equations and conditions appropriate for the state that is eventually reached show flows extending far away, and a magnetic field that stretches out with them.

As material flows out its rotation rate decreases, due to the tendency of the angular momentum to be conserved. That tendency is not totally overcome by the magnetic field's action to keep the material rotating at a rate close to the star's rotation rate. This lowering of the rotation rate results in the magnetic field spiralling around the rotation axis of the star. The tension in the magnetic field due to this spiralling restricts the flow's move-

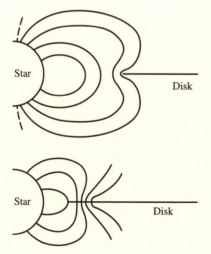

FIGURE 5-11 A Rather Fictional Account of the Evolution of Magnetic Fields near a Rotating LMYSO. The configuration in the lower panel is thought to be relatively realistic for the primary era of mass loss. The disk is indicated by a straight line. Magnetic field lines are curved.

ment away from, but not along, the axis around which the star rotates. Thus, magnetized winds driven by rotation are collimated or channelled outflows.

5.7 Observations of Collimated Outflows and Jets
Around Low-Mass Young Stellar Objects

In the following discussion we will use the term *collimation factor* to indicate the length-to-width ratio. Outflows having collimation factors in the range of 2 to 5 have been observed in CO emission around many young stellar objects (YSOs) of a wide variety of masses. The typical speeds relative to the stars they surround are about 20 km s^{-1}. Most of these outflows are bipolar with lobes on either side of a central star. An object named L1551 is one whose CO outflow has been extensively observed. It has passed through at least four periods of major mass loss. Each lasted about several tens of thousands of years and was separated from its predecessor by a similar period of time. The mass associated with the molecules in the dynamical features caused by each of the mass loss events in L1551 is a few tenths to a few solar masses. The total spatial extent is several light years.

The molecular component in each of these roughly 20 km s^{-1} outflows observed in CO emission is probably mostly gas swept up from the molecular region in which the star formed. The mass of the material actually lost from the star is probably smaller.

Attempts to observe the winds of the LMYSOs, rather than the material that the winds have swept up, have been successful in some cases. Observations of 21 cm emission from neutral atomic hydrogen flowing from young stars have shown the neutral atomic wind of each to be moving at roughly 200 km s^{-1} and the mass loss rate of each to be between a millionth to ten millionths of a solar mass per year. Such a high mass loss rate from a solar mass YSO can be sustained for only tens of thousands to several hundred thousand years.

The highly collimated outflows detected in optical and radio emission from gas at temperatures around 10,000 K are spectacular in some cases. Plate 4 shows an optical image of the jet of the so-called HH34 system. It extends up to about five light years. Many jets are about a factor of 10 shorter. The speeds of gas in the jets are several hundred kilometers per second.

The HH in HH34 refers to Herbig-Haro, a combination of the surnames of the two scientists who, early in the 1950s, discussed knots of optical emission with number densities and temperatures of about 10,000 cm^{-3} and 10,000 K in regions containing young stars. The presence of such knots is apparent in plate 4, but they are also observed in regions for which optical photographs do not reveal such clearly defined jets, possibly as a consequence of obscuration of the jets by the dust in surrounding molecular gas. The

Herbig-Haro objects, as the knots are called, probably form in some cases as consequences of the development of two-shock structures (similar to that depicted in figure 4-6 but limited in spatial size by the highly collimated nature of the jets) as jets interact with the surrounding cloud material. Internal reflections of the fast gas can occur in jets, and such reflections may drive shocks that may be the cause of some Herbig-Haro object formation. Internal reflections may be responsible for the existence of several Herbig-Haro objects along a jet, but separate outbursts in the jet source may be responsible as well.

The use of an array of millimeter-wave telescopes, acting together as one, reveals in some sources the presence of rapid but relatively cool molecular flows collimated nearly like the optical jets are. The speed of such a highly collimated molecular outflow may be as high as 100 km s^{-1}. The fast molecular material may be ambient material entrained in the fast jet flow by friction at the jet-ambient material interface. Fast but warmer molecular flows associated with jets and Herbig-Haro objects are observed in infrared emission of H_2 at wavelengths of about 2 microns. The emission is due to transitions between different vibrational states of the H_2 (cf. section 6.4). Such H_2 emission figures prominently in the study of outflows interacting with ambient material in regions of high-mass star formation, treated in the next chapter.

SELECTED REFERENCES

Bachiller, R. "Bipolar Molecular Outflows from Young Stars and Protostars." *Annual Review of Astronomy and Astrophysics* 34 (1996): 111–154.

Cowie, L. L. "Cloud Fluid Compression and Softening in Spiral Arms and the Formation of Giant Molecular Cloud Complexes." *Astrophysical Journal* 245 (1981): 66–71.

Dyson, J. E., and Williams, D. A. *The Physics of the Interstellar Medium.* 2nd ed. Bristol: Institute of Physics Publishing, 1997.

Hartquist, T. W., and Williams, D. A. *The Chemically Controlled Cosmos.* Cambridge: Cambridge University Press, 1995.

Hartquist, T. W., Caselli, P, Rawlings, J. M. C., Ruffle, D. P., and Williams, D. A. "The Chemistry of Star Forming Regions." In *The Molecular Astrophysics of Stars and Galaxies*, ed. T. W. Hartquist and D. A. Williams, 101–137. Oxford: Clarendon Press, 1998.

Ray, T. P. "Early Days in the Life of a Star." *Scientific American* 283 (2000): 30–35.

Shu, F. H., Najita, J., Ostriker, E., and Shang, H. "Magnetocentrifugally Driven Flows from Young Stars and Disks V Asymptotic Collimation into Jets." *Astrophysical Journal* 455 (1995): L155–L158.

6

REGIONS OF HIGH-MASS
STAR FORMATION

A high-mass star contains more than about eight solar masses of material. Regions where such stars form differ significantly from those in which only low-mass stars form. High-mass stars are thought to be born in clumps that are magnetically supercritical, whereas clumps in which only low-mass stars appear are probably magnetically subcritical (cf. section 5.3). High-mass stars have more powerful winds than low-mass stars and they evolve into supernovae; consequently they affect much larger volumes of the environments surrounding them. Low-mass stars emit negligible amounts of ultraviolet radiation during their youth and main-sequence phases, whereas high-mass stars ionize large amounts of material and drive motions with the ultraviolet radiation that they emit.

The ionizing radiation of a high-mass star creates a source called an H II region. (H II is the notation used by spectroscopists when referring to the proton or hydrogen ion, H^+). Some of the smaller H II regions are classified as *ultracompact*; some larger ones are called *classical*.

Masers, sources that are the longer wavelength relatives of lasers, are found around some ultracompact H II regions. They are seen in emission lines of molecules, including OH (hydroxyl), CH_3OH (methanol), and H_2O. The chemistry in them is influenced by the stellar radiation fields.

Molecular clumps called *hot cores* are also found in the proximity of ultracompact H II regions. They are heated by the hot stars' radiation fields to temperatures of 100 degrees or more and typically contain at least 100 solar masses of material.

The Orion Nebula is a nearby region of high-mass star formation. Infrared emission from H_2 has been observed to study the shocks that develop in the wind-ambient medium interaction near that region. Shocks in lowly ionized, magnetized molecular material have more complicated structures than shocks in moderately or highly ionized gas. High spatial resolution, as well as high spectral resolution, data for the H_2 emission have shown that theoretical understanding of the wind-ambient medium interaction is incomplete.

That interaction may actually induce sequential high-mass star formation and influence low-mass stellar birth.

6.1 H II Regions

Throughout much of this volume we address bubbles caused by mass loss from various types of objects. Bubbles are also blown by the interaction of radiation with matter.

In the vicinity of a young or main-sequence high-mass star, the absorption of ultraviolet radiation by neutral atoms forms charged ions and unbound electrons. The process of ionization due to photoabsorption is called photoionization, and it heats gas near a high-mass star at a sufficient rate to maintain the temperature at about 10,000 K.

The transition zone between the region where hydrogen is nearly all ionized to where it is nearly all neutral is thin compared to the size of the H II region. Most of the neutral material will typically have a temperature of roughly 100 K or less. The large temperature (and pressure) difference means that H II regions expand until the pressure in them equals that of the surrounding neutral gas.

In 1954 the outstanding astrophysical gas dynamicist Franz Kahn, who supervised the thesis research of one of the authors (J. E. D.), published the seminal paper on the theory of the dynamics of H II regions. Following Kahn, we assume that the radiation turns on either at the same time or well before the stellar wind does, and that initially the ambient medium is uniform and neutral. These assumptions allow us to simplify the presentation without losing the key points of H II region evolution.

A region about a light year in extent is ionized in about 100 years. To calculate the radius of the region that is ionized so rapidly, we must understand the concept of hydrogen recombination. It is the process of an electron being captured by a hydrogen ion to create a neutral atom. After recombination produces a neutral atom, it is again photoionized. Rapid expansion of the H II region occurs until its volume is large enough for the number of recombinations occurring per unit time in it to be comparable to the number of ionizing photons emitted by the star per unit time. (We have used "comparable" in the previous sentence because about 30 percent of recombinations lead to the production of photons that can induce further ionization, thus affecting the H II region size.)

Just when the phase of most rapid expansion concludes, the density of the ionized material and the neutral material around it are about equal, but the pressures of the two regions differ by a factor of about 100 or more, due to the temperature difference. The hotter gas expands at a speed comparable to its sound speed of about 10 km s^{-1}, driving a shock into the neutral material. As the density of the hot gas drops, the recombination rate per particle drops, and ionizing photons can travel further and ionize more gas. The speed of the expansion of the hot gas drops with time, and the shock eventually fades away. Expansion continues until the pressure of the hot gas is that of the neutral surrounding material.

The ultimate size of an H II region depends on the density of the surrounding medium and the ultraviolet power of the star. H II regions of different sizes are given different names. If the star is surrounded by the gas of an extended neutral hydrogen cloud with a number density of about 100 per cubic centimeter, the H II region will be tens of light years across; an H II region of this size is referred to as a *classical H II region*. Plate 5 depicts a classical H II region. If the high-mass star is surrounded by gas having a number density of 10 million neutrals per cubic centimeter, which is typical of dense cores associated with high-mass star formation, the size of the H II region is several tenths of a light year. An H II region of this size is said to be an *ultracompact H II region* or UCHIIR.

Intermediate size regions are called *compact* H II regions; much smaller ones are said to be *hypercompact*. The largest H II regions are observed in distant galaxies, where they are excited by thousands of stars and are many thousands of light years in extent.

The photons ionize other atoms as well as hydrogen. Much of the observed optical emission is from the ions O$^+$, O^{++}, and N$^+$. The emission is induced when an unbound electron collides with an ion and causes a bound electron in it to be excited to a more energetic bound state; the more energetic state depopulates itself through the emission of a photon. This mechanism is discussed again in section 8.4. Other observed optical emission occurs in the recombination of H$^+$ with an electron to form H.

The H II regions also emit radio waves with wavelengths of centimeters to tens of centimeters. The continuum radio emission arises in the scattering of unbound electrons with ions and is called *free-free* emission or *bremsstrahlung*, which is considered again in section 8.4. Many UCHIIRs were discovered only by the detection of their radio emission, because they are often very deeply embedded in molecular clouds where the dust content is high enough to absorb emission in the visible or shorter wavelength regions of the spectrum.

6.2 Molecular Masers

Rather intriguingly, masers are found around UCHIIRs. As stated in the introduction to this chapter, masers are related to lasers, and the word *maser* is an acronym for "microwave amplification of stimulated emission of radiation," whereas the *l* in *laser* indicates "light." Maser emissions arise in molecules surrounding UCHIIRs. Figure 6-1 shows the locations on the sky of OH (hydroxyl) and CH_3OH (methanol) masers with respect to the continuum radio emission of an UCHIIR. The maser emissions are at discrete wavelengths between 2.4 cm and 18 cm.

To understand what the maser emission is, one must first review the natures of molecules, upon which we touched briefly in section 5.1. Many readers will be more familiar with atoms. In an atom an electron that is bound can have only specific discrete energies characteristic of the type of atom. In a molecule in which the different atoms are bound together, the vibration within and the rotation as a whole of the molecule can have only discrete energies associated with them; these allowed energies of vibration and of

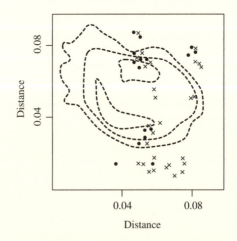

FIGURE 6-1   Masers Located Near an Ultracompact H II Region. Filled circles show the locations of OH masers and closed circles indicate those of CH_3OH masers. Dotted curves are contour plots of the radio continuum emission from the UCHIIR in the source W3(OH). The continuum emission is strongest in regions surrounded by the most contours. Figure adapted from Menten, K. M., Reid, M. J., Pratap, P., Moran, J. M., and Wilson, T. L., "VLBI Observations of the 6.7 GHz Methanol Masers Toward W3(OH)," *Astrophysical Journal* 401 (1992): L39–L42.

rotation vary from one type of molecule to another. The energy difference between two low-lying rotational levels or states of a nonvibrating molecule are typically much less than the energy difference between different bound vibrational states. The methanol maser emission arises in transitions between rotational states of nonvibrating molecules.

For some types of molecules, a rotational state in a molecule is split into sublevels. The energies of two such sublevels in a hydroxyl molecule differ, because in the different sublevels various magnetic moments interact differently. Some of the moments are caused by the electric currents associated with the electron orbits, whereas others are due to electrons or nucleons spinning about their own axes.

Now we turn to what it means for a transition to show maser (or laser) activity. Consider two separate levels or states in each of a group of identical molecules and suppose that the molecules are in a dark environment. Collisions between two molecules can induce excitation of one or both molecules to the higher energy state. Further collisions can cause de-excitation of the high-energy state. The emission of photons can also cause de-excitation. If the density is low, collisions will be rare, and radiative de-excitation of the molecules will keep the population of the upper state small. If the density is high enough collisions are rapid, and the population of the upper state relative to that of the lower state depends only on the temperature of the gas. A low enough temperature will cause a low population of the upper state, whereas a high enough temperature will cause comparable populations in the upper and lower states. For dense gas there is a one-to-one relationship between the ratio of the populations of the upper state to the lower state and the temperature. Masing (or lasing) occurs when the ratio of populations exceeds that given by this one-to-one relationship. Thus, more radiation is emitted in the transition between the two states than in normal circumstances. The radiation induces or stimulates further radiation when it interacts with other molecules in the upper state. That is, when a photon emitted by the decay of an excited state encounters another molecule in the same excited state, the photon can cause the emission of a further photon. An avalanche of photons occurs in the maser or laser process. The methanol and hydroxyl masers appear tens of billions of times (and more) brighter over the narrow wavelengths at which they are observed than would be expected if conditions were not favorable for the masering to occur!

The attainment of the high population of the upper of the two states, necessary for masing to occur, requires that a third state must be involved in the process. An example of how the involvement of the third state might create a high population of the upper state is depicted in figure 6-2. It shows schematically the collisionally induced excitation of the third state, which has the highest energy. In some types of molecules the highest state radi-

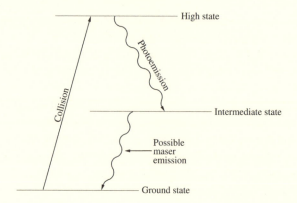

FIGURE 6-2   A Model of the Collisionally Induced Pumping of a Maser. Collisions of the molecule with another molecule cause the excitation of a state with an energy above that of the ground state. Photoemission causes the decay of the state to populate preferentially a state with an energy intermediate between that of the highest state and the lowest state. A sufficiently high population of the intermediate state due to the preferential decay to it will create conditions favorable for masing.

ates rapidly to populate preferentially the more energetic of the other two states. This is said to be a *collisionally pumped system*. There are also radiatively pumped systems in which the excitation from the lowest of the three states to the highest is due to the absorption of radiation, but the radiative decay is preferentially to the higher of the two states.

Hydroxyl masers are used to measure magnetic field strengths in the vicinities of young stars, as the OH sublevels' energies are affected by the magnetic field. Observations of water masers revealed early evidence for motions at speeds as high as about 100 km s^{-1} in some star forming regions, even though slower water masers are also seen. Molecular masers are mentioned in later chapters in the context of a variety of types of sources.

6.3 Hot Cores

The molecular clumps called *hot cores* are often found associated with UCHIIRs. The hot cores' temperatures of about one to several hundred degrees are much higher than those of dense cores in regions of low-mass star formation. Their masses and number densities range up to several hundred solar masses and around 100 million per cubic centimeter, respectively. Thus, they are more massive and appear to be denser than the dense cores in regions of low-mass star formation.

The hot cores are heated by dust in them absorbing infrared radiation emitted by dust nearer to the high-mass stars in their vicinities. This dust nearer to the stars is heated, by the absorption of the starlight, to high enough temperatures to emit the infrared radiation.

The heating of the dust in the hot cores causes them to have particularly rich molecular compositions. The heating evaporates ices that formed on the dust surfaces when the regions were colder. Some of the molecules released into the gas phase were formed by chemistry that was catalysed by the dust. For instance, methanol is produced primarily on dust surfaces, and the methanol that is detected in maser emission (as described in the previous section) was released from grains as a consequence of high-mass stars being born.

As time passes, the chemicals released from the dust surfaces react with one another, and the molecular composition alters. Models of the chemistry in hot cores have been developed in order to facilitate the inference of how long ago the hot cores were first heated. Thus, the chemistry is used as a clock, which is possible because it is evolving from a state influenced by the "recent" injection of material from the surfaces to one dominated by the reactions occurring in the gas phase. Hot cores are estimated to have been initially heated around several tens of thousands to about 100 thousand years ago. Efforts are in progress to determine whether the chemistry can be used to infer whether hot cores have been shocked by interactions with the winds of young stars before, or somewhat after, the heating by infrared radiation became important. If successful, these efforts may reveal information about how the winds and radiation fields of young high-mass stars turned on.

6.4 The First Studies of the H_2 Emission from the Orion Star Forming Region

The Orion Molecular Cloud Complex is a giant molecular cloud at a distance of about 1,400 light years and contains several regions of ongoing or recent high-mass star formation. The Orion Nebula, one of the most spectacular objects visible from the Earth, is pictured in plate 6. It is a compact H II region having a diameter of about 2 light years, small compared to the total extent of the Orion Molecular Cloud Complex, which is hundreds of light years across.

The Becklin-Neugebauer-Kleinmann-Low (BN-KL) infrared nebula lies about a half light year nearly directly behind the Orion Nebula. It is illuminated primarily by a star that is younger than those ionizing and heating the Orion Nebula. The dominant star in the BN-KL region has formed so recently that its wind and radiation field have not yet cleared away much of the material left over from its birth. This remnant placental material obscures the star, is heated by its radiation, and reemits the energy at infrared wavelengths of about a micron and longer.

Star formation may well be proceeding sequentially from the Orion Nebula into the molecular material behind it, creating a deepening blister in this region of the giant molecular cloud. The sweeping up of material around the bubbles blown by young high-mass stars seems to induce further star formation, at least in this region!

The BN-KL nebula is amongst the most observed objects in the Universe. In the late 1970s, an infrared spectrometer was used to search for H_2 emission toward it. Emission was found in the search, and figure 6-3 shows contours of its strength as measured in those early observations. The emission comes from a region extending about a third of a light year. It is concentrated in distinct narrow features arising due to the radiative depopulation of H_2 in excited vibrational levels. In an excited vibrational level, the nuclei in the molecule move back and forth toward one another much as they would if attached to one another by springs. This vibrational motion can be excited in collisions between molecules if the temperature is roughly 1,000 K or higher. Much of the emission arises from the depopulation of levels that are excited rotationally as well as vibrationally. The nuclei in an H_2 molecule can rotate about an axis midway between them and perpendicular to the line passing through them both. The rotational motion can also be excited in collisions between molecules if the temperature is high enough. Temperatures sufficient for collisions to induce even the lowest vibrational excitation are sufficient for them to cause excitation to a half dozen or more excited

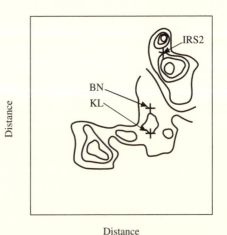

Distance

FIGURE 6-3 H_2 Emission from the BN-KL Nebula as Measured in the 1970s. Contours of the intensity of an H_2 emission feature are shown. The locations of peaks in infrared emission from dust are indicated by BN, KL, and IRS2. Figure adapted from Beckwith, S., Persson, S. E., Neugebauer, G., and Becklin, E. E., "Observations of the Molecular Hydrogen Emission of Orion," *Astrophysical Journal* 223 (1978): 464–470.

rotational levels. Heating of the gas can arise due to the interaction of a young star's wind or outflow with ambient molecular gas, or by the absorption of stellar radiation.

6.5 Magnetic Moderation of Flows in Molecular Regions

The first observations of H_2 emission in the BN-KL nebula were of low angular resolution and also of low spectral resolution by present standards. Of course, low angular resolution means that the fine spatial structure of an object gets washed out in an image of it. Low spectral resolution means that a photon registered by the detector can have any wavelength (or frequency) in a range that is wide compared to the range over which the brightness versus wavelength (or frequency) profile of a spectral feature might be expected to show interesting subfeatures. Despite the low resolution of the first H_2 emission observations, they showed clearly that some of the H_2 emitting gas has speeds of about 100 km s^{-1}. This is far higher than the roughly 24 km s^{-1} that is the maximum speed of a shock in a dense nonmagnetized medium through which H_2 can pass and survive. A shock of too high a speed would raise the temperature to such an extent that collisions between the molecules could break them apart.

As a consequence of the large speeds associated with the H_2 gas in the BN-KL nebula, theoreticians began to focus attention on how magnetic fields in a molecular region might moderate the destructive effects of a fast shock. The moderating process operates because ions can move relative to neutrals in a magnetized medium containing many fewer ions than neutrals. Such relative motion of ions with respect to neutrals was described in section 5.5.

To see how the relative motion of ions and neutrals can affect a shock, consider the motion of a large object in a direction perpendicular to the magnetic field lines in the material ahead of it.

If there were absolutely no collisions between charged particles and neutrals, we could treat the charged particles as one separate fluid and the neutrals as another. As described before, the presence of the magnetic field affects the nature of, and speed of, sonic waves in the ion fluid (cf. the last paragraph of section 5.2). For a given magnetic field strength, the fast magnetosonic speed is higher the lower the mass density of ions. Thus, in a region of low ionization in a star forming region, the speed of a sonic wave in the ion fluid is much higher than the sound speed in the neutral fluid.

Consequently, a large object moving at a speed much higher than the sound speed in the neutral fluid may be moving at a speed much lower than the sonic speed in the ion fluid. Then the ion fluid has the ability to adjust its distribution in front of the oncoming object rather than experience a shock. If there were no collisions between ions and neutrals, the neutral fluid would experience an ordinary hydrodynamic shock, like those addressed in section 2.1.

In reality, as depicted in figure 6-4, the existence of collisions between charged particles and neutrals results in the information carried upstream ahead of an oncoming object by the ion fluid being partially transmitted to the neutral fluid. This transmission of information results in a distortion of the distribution of the neutral fluid ahead of the object that, in some cases, prevents a well-defined shock in the neutrals from occurring. However, the

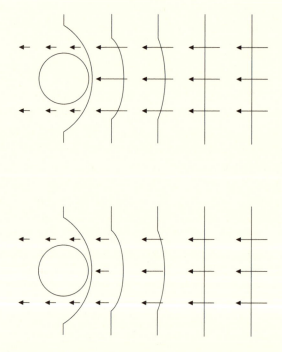

FIGURE 6-4   Magnetic Moderation of Flow in a Medium with Few Ions. A view of the velocity of neutral gas and of the magnetic field lines from the perspective of an object if no collisions occur between neutrals and charged particles is shown in the upper half of the figure. Solid lines depict magnetic field lines which, due to the object's presence, begin to bend far away from the object. The distant bending is a consequence of the relative speed between the object and the flow being below that of the fast magnetosonic waves in the charged particle fluid. Arrows show the velocity of the neutral flow at various points. It is undisturbed until it passes through a shock near the object.

A view of the velocity of the neutral gas and of the magnetic field lines if there is moderate collisional coupling (friction) between the charged particles and the neutrals is shown in the lower half of the figure. The magnetic field structure is somewhat less distorted far from the object than is shown previously. The neutral flow velocity varies smoothly and no sharp shock exists.

1 An Aurora Viewed from Space. The aurora occurred in the southern hemisphere and consequently is referred to as an Aurora Australis. Notice the crown feature. Image courtesy of NASA.

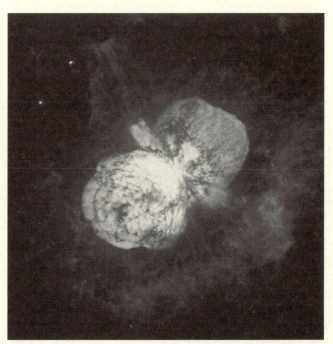

2 The Outflow of an
Evolved Massive Star
(eta Carina). Jon Morse
kindly provided the
image. Image courtesy
of Jon Morse (University
of Colorado), Kris
Davidson (University of
Minnesota), and NASA.

3 An Optical Image of
a Planetary Nebula
(NGC 2392). This
source is often called
the Eskimo Nebula. It
is roughly a light year
across. Image courtesy
of NASA, A. Fruchter,
ERO Team, and STScI.

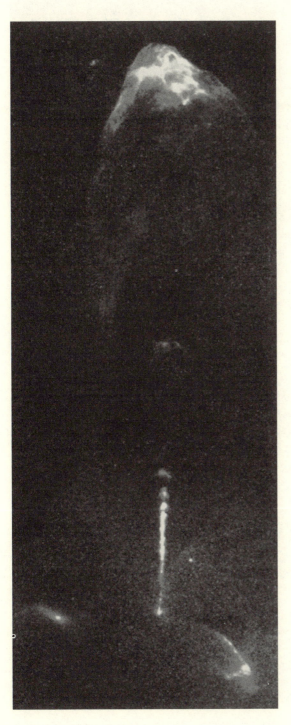

4 The Jet of HH34.
Bo Reipurth kindly
provided the image.

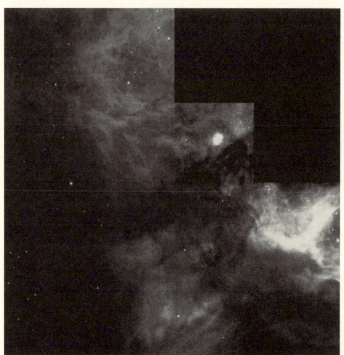

5 A Classical HII
  Region (N 159).
  Image courtesy
  of NASA/STScI.

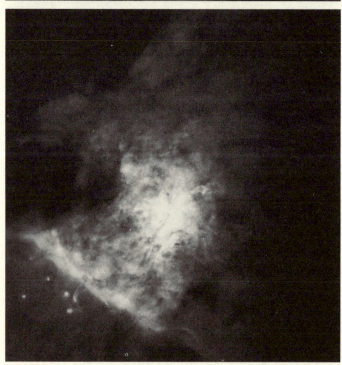

6 A Mosaic of the Orion Nebula. The image was generously provided
  by C. R. O'Dell. Image courtesy of C. R. O'Dell (Vanderbilt
  University), S.-K. Wong (Center for Astrophysics), and NASA.

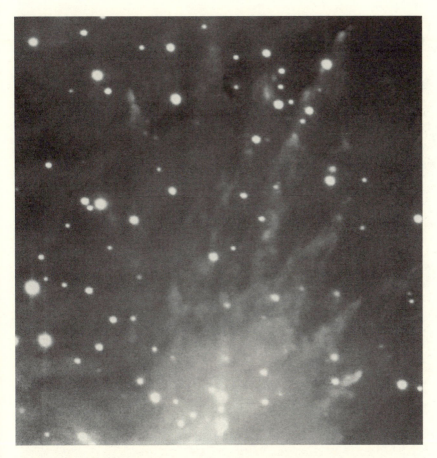

7 Molecular Bullets in the Orion BN-KL Region. The image was produced through the overlaying of three separate "false color" images using H_2 emission, dust emissions, and ionized iron emission. Michael Burton generously supplied the image. The observations are described in Allen, D. A., and Burton, M. G. "Explosive Ejection of Matter Associated with Star Formation in Orion Nebula." *Nature* 363 (1993): 54–56.

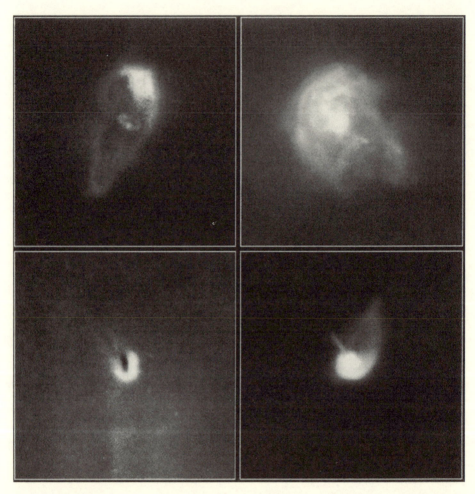

8 Proplyds in Orion. The images show the variance in proplyd structures. Each proplyd is somewhat larger than the solar system. John Bally kindly provided the image. Image courtesy of NASA, J. Bally (University of Colorado), H. Throop (SWRI), and C. R. O'Dell (Vanderbilt University).

9 A Seyfert Galaxy (NGC 7742). The radius of the ring is about 3,000 light years.
Image courtesy of Hubble Heritage Team (AURA/STScI/NASA).

10 Jets of an AGN (3C 288). The distribution of radio emission with a wavelength of 4 centimeters is shown. Jets and the radio lobes that they power are clearly seen. The total extent of the emission is about 300,000 light years. The image was kindly provided by A. H. Bridle. Image courtesy of A. H. Bridle, J. A. Callcut, E. B. Fomalont, and NRAO/AUI.

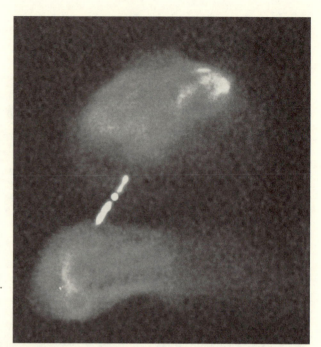

11 The Southern Crab (He2–104). The nebula is several light years long. Hugo Schwarz kindly provided the image. Image courtesy of R. Corradi (Instituto de Astrofisica de Canaries), M. Livio (STScI), U. Munari (Osservatorio Astrofisico di Padova-Asiago), H. Schwarz (Cerro Tololo Inter-American Observatory), and NASA.

collisions between ions and neutrals result in frictional heating. Thus, the neutrals, even in a magnetically moderated flow, can reach temperatures between about 1,000 and about 4,500 K, which typify regions of substantial H_2 infrared emission.

The effects of magnetic moderation do not remain significant for arbitrarily high speeds. For standard assumptions about magnetic fields in star forming regions, only disturbances slower than 40 km s^{-1} will leave molecular hydrogen intact; 40 km s^{-1} is higher than the speed limit for the survival of molecular hydrogen in a shock that is not moderated by a magnetic field, but it is still considerably lower than the speed associated with the H_2 that gives rise to the emission observed in the BN-KL region.

One suggestion offered to solve this problem is that the magnetic field strength in some of the BN-KL region's material is much higher than previously expected.

6.6 Later High-Resolution Studies of the H_2 Emission in the Orion Star Forming Region

In astronomy, low angular resolution and low spectral (or wavelength) resolution observations are generally the first type made to study a source. This is usually a consequence of the simple fact that, typically, significant technological advances are required to develop instruments capable of providing higher resolution data for the distant objects that astronomers study. Although lower resolution observations can unveil interesting information and stimulate major theoretical investigations, higher resolution results often present problems that differ qualitatively from those suggested by the lower resolution ones. The detection of H_2 infrared emission from the Orion BN-KL region at higher angular and spectral resolutions provides an excellent example of how even more detailed data alter the nature of questions raised. The dedicated attack on the region was spearheaded by Peter Brand of the University of Edinburgh, Michael Burton of the University of New South Wales, and their collaborators. Professor Brand has also invented imaginative theoretical models to explain the data.

The lower resolution studies of the Orion BN-KL region, described in section 6.4, led to the recognition of the interesting problem concerning how H_2 can reach speeds much higher than those at which it is destroyed in shocks (unless an unexpectedly strong magnetic field moderates the shock). The strongest emission is from the vicinity of a source referred to as IRS 2 or IRC 2 of the BN-KL region (cf. figure 6-3).

A higher angular resolution, low spectral resolution map of the H_2 emission was obtained for the vicinity of IRS 2. Maps have also been made in continuum infrared emission from the dust (which is heated by the stellar radiation) and in emission from singly ionized iron atoms. Plate 7 is an image composed of different emissions. Iron emission is seen at the heads of what are apparently bullets of neutral gas ejected from IRS 2. A long wake is detected in H_2 emission around each of the bullets; it points back toward IRS 2. That the interaction of a group of bullets (rather than a more uniform wind) with the ambient material is responsible for generating the H_2 emission came as somewhat of a surprise when this higher angular resolution data first became available.

Subsequently, H_2 and singly ionized iron emission data obtained at an angular resolution comparable to that of the image in plate 7, but with much higher spectral resolution, were obtained. Researchers had expected that each H_2 wake would show emission in two broad but clearly separated velocity ranges, just as in the wake around a boat some water has a velocity with a component to the port side and some to the starboard side. Instead the H_2 emission in a bullet wake is found to have a single broad velocity feature, presenting a puzzle that presently remains unsolved.

6.7 Cometary-Like Tails Near the Orion Nebula

Objects referred to as *proplyds* are distributed around the Orion Nebula at distances from its center several times larger than the linear extent of the region represented in plate 6. (The term *proplyd* comes from *pro*to-*pl*anetary *d*isks.) Plate 8 shows images of just a few of the many tens of proplyds discovered in the vicinity of the Orion Nebula. It is striking that each of the proplyds possesses a tail, and except in rare cases (in which the proplyds are near one another) the tails point almost directly away from the brightest star in the Orion Nebula.

Many of the proplyds are believed to be disks around low-mass young stellar objects. Hence, the vicinity around the Orion Nebula is one in which copious formation of low-mass stars as well as of high-mass stars is occurring.

Two explanations have been offered for the development of the thin tails of the proplyds. One is that such a tail arises due to the shadowing (by the main body of a disk) of the tail from ionizing radiation coming from the hot stars in the Orion Nebula. In the other picture, the flow of a mixture of stellar wind material and of material evaporated from other clumps encountered by the wind acts to form the tail. The flow must be at a speed not much in excess of the sound speed in the mixture or less. The impact of a highly supersonic wind directly on a clump would cause it to have a thermal pressure much higher than the thermal pressure of the wind surrounding the clump. This large pressure difference would result in evaporation over a broad downwind zone. In contrast, if the flow is only just faster than the sound speed or slower, the thermal pressure of the clump is not driven to a high value relative to the pressure of the surrounding wind. Then the wind's pressure prevents substantial expansion of evaporating clump material, except in the direction of the undisturbed wind's flow.

We consider other cometary-like tails in the last section of the next chapter.

SELECTED REFERENCES

Allen, D. A., and Burton, M. G. "Explosive Ejection of Matter Associated with Star Formation in the Orion Nebula." *Nature* 363 (1993): 54–56.

Bally, J., O'Dell, C. R., and McCaughren, M. J. "Disks, Windblown Bubbles, and Outflows in the Orion Nebula." *Astronomical Journal* 119 (2000): 2919–2959.

Dyson, J. E., and Franco, J. "H II Regions." In *Encyclopedia of Astronomy and Astrophysics*, ed. P. Murdin, 1114–1119. Bristol and London: Institute of Physics Publishing and Nature Publishing Group, 2001.

Dyson, J. E., Hartquist, T. W., and Biro, S. "Mass-Loaded Astronomical Flows V.-Tails: Intermediate-Scale Structures in Flowing Clumpy Media." *Monthly Notices of the Royal Astronomical Society* 261 (1993): 430–434.

Hartquist, T. W., and Caselli, P. "Shock Chemistry." In *The Molecular Astrophysics of Stars and Galaxies*, ed. T. W. Hartquist and D. A. Williams, 179–199. Oxford: Clarendon Press, 1998.

Menten, K. M., Reid, M. J., Pratap, P., Moran, J. M., and Wilson, T. L. "VLBI Observations of the 6.7 GHz Methanol Masers Toward W3(OH)." *Astrophysical Journal* 401 (1992): L39–L42.

Pavlakis, K. G., Williams, R. J. R., Dyson, J. E., Falle, S.A.E.G., and Hartquist, T. W. "The Modification by Diffuse Radiation of 'Cometary Tail' Formation Behind Globules." *Astronomy and Astrophysics* 369 (2001): 263–268.

Tedds, J. A., Brand, P.W.J.L., and Burton, M. G. "Shocked H_2 and Fe^+ Dynamics in the Orion Bullets." *Monthly Notices of the Royal Astronomical Society* 307 (1999): 337–356.

Viti, S., Caselli, P., Hartquist, T. W., and Williams, D. A. "Chemical Signatures of Shocks in Hot Cores." *Astronomy and Astrophysics* 370 (2001): 1017–1025.

7

WINDS FROM MAIN-SEQUENCE AND POST-MAIN-SEQUENCE STARS

Winds of LMYSOs are driven by the rotation of magnetized material (cf. section 5.6). A number of other mechanisms can also drive mass loss. This chapter contains treatments of the mechanisms causing mass loss from stars that have evolved beyond infancy.

The concept of the escape speed was used in section 4.2. The square of the escape speed from a body is approximately equal to the product of its mass with the universal gravitational constant, G, divided by the distance from the center of the object. Generally the mechanisms driving mass loss from objects result in outflows having speeds of about two to three times the escape speeds from the objects' surfaces. Thus, the mechanisms are fairly energy efficient in the sense that each uses only a few times more than the minimum energy required to cause material to escape.

One can usually infer the nature of an outflow's source from a measurement of the outflow's speed. Suppose, for example, that material is moving at 450 km s^{-1} (cf. section 4.4 in which the solar wind's speed is stated to have this value). Then one would infer that the escape speed from the surface of the source is roughly a few hundred kilometers per second, which is comparable to the escape speed from the Sun. Similarly, if a flow is moving at nearly the speed of light, one would infer that it originates in the vicinity of a black hole (cf. section 10.3) where the escape speed is about the speed of light.

7.1 Winds Driven by Thermal Pressure

In section 4.2 we touched upon winds driven by thermal pressure. The solar wind is such a wind. Thermal pressure can drive a wind from a spherical object because the pressure causes the gas of the object to have an extended spatial distribution and because at ever greater distances from the object's center the escape speed decreases. Thus, at large enough distances, the gas extending from an astronomical source will have a sound speed that exceeds the escape speed sufficiently for that gas to flow away as a wind. At such distances, if the flow of the gas is not already supersonic, the force due to the gas's pressure accelerates the material to supersonic speeds.

As was mentioned in sections 4.2 and 4.4, the wind from the Sun is much hotter than the roughly 6,000 K associated with the Sun's surface material emitting yellowish light. The solar wind is composed of matter from the Sun's corona. The word *corona* has a Greek root meaning "crown"; the Sun's corona appears like a fiery crown around the dark surface of the Moon during a total solar eclipse. Gas at the 2 million degrees Kelvin associated with the corona drops in density much less rapidly as it extends above the Sun than the roughly 6,000 K gas does. If the Sun had no corona, its wind would be far more feeble than it actually is.

The production and heating of the Sun's corona remain challenging problems. The Sun's rotation rate varies with the depth into the Sun. This *differential rotation* combined with convection (cf. section 3.1) produces magnetized loops. The magnetic pressure of such a loop causes it to have a lower thermal pressure and lower density than surrounding material. The loop rises due to buoyancy, just as a helium-filled balloon rises in the Earth's atmosphere. Most of the loop emerges above most of the 6,000 degree Kelvin gas but remains anchored below the surface of the Sun. In a process called *magnetic reconnection* the loop releases the energy stored in its magnetic field to heat gas to millions of degrees. The detailed nature of the magnetic reconnection process is not understood. What is certain is that under some circumstances in the Sun's corona, when parts of loops having magnetic fields in different directions contact one another, much of the magnetic energy is released rapidly enough to cause the high temperatures.

The solar wind carries only roughly 10^{-15} to 10^{-14} solar masses per year. The Sun's current mass loss rate is small compared to that of a LMYSO even when its wind is at its most powerful (cf. section 5.7). However, the speeds of a LMYSO wind and the solar wind differ by only a factor of two or so, reflecting the similarity of a LMYSO's and the Sun's escape speeds.

The winds of low-mass, main-sequence stars are thought to be driven by thermal pressure and to have low associated mass loss rates, comparable to that of the Sun.

7.2 Winds from Hot Giants and Supergiants

The thermally driven winds described above are produced because thermal pressure gradients accelerate the gas to speeds high enough for it to escape from the stars. Forces other than that associated with thermal pressure exist; some of them generate winds as well.

In the 1960s, ultraviolet observations, made with rocket-borne instruments, of young hot massive stars showed that they all have spectral lines that show the P Cygni characteristics described in section 1.3. The final wind speed can be determined from the shortest wavelength to which the absorption feature extends. The final speeds measured for these stars range from hundreds to thousands of kilometers per second. The wind speeds are between two and three times the estimated escape speeds. However, the temperatures in the atmospheres of these stars are never more than about 100,000 K (corresponding to a sound speed of about 30 km s^{-1}). These stars do not possess coronae analogous to the Sun's. A sound speed of 30 km s^{-1} cannot produce a powerful thermally driven wind with a terminal speed of about 1,000 km s^{-1}.

The clue to what drives these winds comes from the observational data. Typical values of the mass loss rates of hot giants and supergiants range from about one tenth of a millionth (10^{-7}) of a solar mass per year to perhaps one hundredth of a thousandth (10^{-5}) of a solar mass per year. The mass loss rate is found to correlate with the radiative luminosity of the star, that is, with the amount of energy radiated by the star per second. The radiation fields of the stars are inferred to be implicated in some way in driving the winds. Photons of light carry momentum, and when light is either absorbed or scattered, momentum is transferred. The laws of mechanics tell us that a rate of change of momentum per unit of time is equivalent to a force. Thus light exerts a force on an absorbing or a scattering surface. The force per unit area due to the light is called radiation pressure. We can measure the momentum output rate in the radiation field of a star by measuring the total energy in the radiation field and dividing the answer by the speed of light. If this momentum is responsible for driving the winds from these stars, then we might expect that the momentum output rate in a wind (i.e., the momentum carried away by the wind per unit time) would be no more than the momentum output rate in the radiation field. Indeed, the wind momentum output rates are found to be generally less than 10 percent or so of the radiation momentum rates; there is certainly enough momentum available in the radiation to drive the winds.

The key question therefore concerns how the radiation momentum is communicated to the gas. The detailed answer lies in the composition and physical structure of the wind gas. The gas is mainly hydrogen but contains elements such as carbon, oxygen, nitrogen, and many others. To provide an example we focus on the element nitrogen, which is usually present as an ion, N^{3+}. This is a nitrogen atom that has lost three electrons (and retained four) as a result of collisions with free electrons or absorption of energetic photons. The energies of levels of N^{3+} in which the four electrons are all bound are distinct, and transitions between them have discrete energies. A high energy level can be reached from a low energy level by the absorption of a photon.

Photons from the stellar radiation field can be absorbed by N^{3+}, and when absorption occurs, the momentum of the photon is transferred to the ion, that is, it is suddenly accelerated. If the ion does not decay from the high excited level to the original level and produce a similar photon, the original photon is lost, resulting in true absorption. However, sometimes a similar photon is produced by decay; the absorption of a photon followed by the emission of a similar one is considered to be a type of scattering. In either absorption or scattering the photon momentum produces acceleration of the ion. An accelerated ion pulls or pushes on its neighbors through its electrical interaction with them, leading the whole gas to accelerate. Acceleration takes place as a result of absorption and scattering by many different types of ions. Hydrogen plays little role in the acceleration, because most of it is in H^+, which does not interact with radiation as efficiently as an ion with bound electrons. The fraction of energy that goes into these winds from the radiation field is small; usually less than one percent or so of the radiation energy is converted into wind energy.

As in the case of thermally driven winds, the winds are gradually accelerated to supersonic speeds. A major difference is that in a radiatively driven wind there can be very significant acceleration even after it has become supersonic. Consequently, radiatively driven winds can have speeds higher compared to the escape speed than winds driven by some other mechanisms.

7.3 Winds from Wolf-Rayet Stars

Wolf-Rayet stars are highly evolved massive stars that have experienced considerable mass loss (cf. section 3.3.2). They are somewhat cooler than young massive stars but have similar luminosities. We might anticipate radiatively driven winds from Wolf-Rayet stars to have mass loss rates and speeds similar to those of younger high-mass stars. Observations do reveal that the maximum speeds of Wolf-Rayet winds are similar to those of the young massive stars, reflecting the fact that these different groups of stars have very similar escape speeds. However, rather surprisingly Wolf-Rayet stars were found to have mass loss rates that are usually about ten times or so higher than those of young massive stars. A very careful comparison of the wind momentum output rates and the momentum output rates of the associated radiation fields showed that in the Wolf-Rayet stars the former are typically about a factor of 10 higher than the latter, the opposite of what is found for young massive star winds.

The resolution of this apparent paradox can be found through a reconsideration of the way in which momentum is transferred to gas by absorption and scattering. If absorption takes place, the photon is lost from the radiation field. If scattering takes place, a new photon is generated; it can also be scattered. Of course, the regenerated photon can be emitted in any direction, even back in the direction of the star. If scattered while moving back toward the star the photon retards the wind. However, usually, the photons tend to diffuse away from the star and the net result of many scatterings per original photon is an outward push on the gas. The key result is that multiple scattering increases the momentum that can be transferred to the gas by roughly a factor of 10 over single scattering or absorption. Multiple scattering seems adequate for the production of the very high Wolf-Rayet mass loss rates.

7.4 The Role of Dust in Mass Loss from Some Evolved Stars

The Wolf-Rayet stars are very hot evolved stars, but lower mass stars in the AGB phase of evolution (cf. section 3.3.2) have very much cooler surfaces, with temperatures of only a few thousand degrees. Molecules are abundant in the cool outer atmospheres of many AGB stars, and the chemistry occurring in the coolest material gives rise to the formation of dust in some of them. Dust formation in an AGB atmosphere containing more carbon than oxygen is probably akin to soot formation in terrestrial acetylene burning and combustion. Dust production in atmospheres with more oxygen than carbon is not so well understood, as most carbon is then in CO, which does not burn, and the dust is formed from silicates and oxides of magnesium and iron. The production of these metallic oxides in "burning" is of less practical terrestrial interest than hydrocarbon soot formation. This lower practical interest in the high-temperature production of metallic oxides has contributed to our lack of a detailed understanding of the stellar production of such material.

Dust produced in the atmospheres of some AGB stars absorbs and scatters the stellar radiation. Consequently, radiation pressure pushes it away from the star. The collisions of dust with gas tends to drag the gas along. Also the dust carries electric charge. The dust's interaction with the magnetic field in a stellar atmosphere contributes to the acceleration of the gas. Thus, radiation pressure on dust can drive mass loss. The AGB stars in which radiation pressure on dust may play a significant role have winds carrying around one hundredth of a thousandth (10^{-5}) of a solar mass of material per year, at speeds of roughly 20 km s^{-1}. These stars are destined to be surrounded by planetary nebulae.

7.5 Winds Driven by Wave Dissipation

As seen in section 7.1, the Sun's thermally driven wind has a very low mass loss rate compared to the mass loss rates of high-mass stars. The loss rates found for massive hot stars are due to the radiative driving mechanism discussed in sections 7.2 and 7.3. Some stellar winds have mass loss rates that are far too high to be thermally driven, yet originate on stars having luminosities so low that the winds cannot be driven by radiation. The low luminosity stars with high mass-loss rates include main-sequence stars roughly twice as massive as the Sun, post-main-sequence giants and supergiants, which are probably too hot to have the dust-driven winds described in section 7.4, and perhaps also low-mass stars that have not yet reached the main-sequence.

The winds of these stars may be driven by waves. The waves could be generated by motions (e.g., convection) occurring beneath the stellar surfaces and may propagate upward in the stellar atmospheres. Damping of the waves in a stellar atmosphere may occur by a variety of processes. The damping of the waves would lead them to transfer all of their momentum to the atmospheres and push gas outward (cf. the last two paragraphs of section 5.3). The stars are magnetized, and therefore the waves could be Alfvén waves (cf. paragraph seven of section 5.2).

7.6 Masers in the Outflows of AGB Stars

The outflows of some AGB stars contain masers (cf. section 6.2). The masers are found only in outflows in which oxygen is more abundant than carbon. If the outflows were too rich in carbon, nearly all of the oxygen would be in CO rather than available to form SiO, H_2O, and OH, in which the transitions giving rise to the observed maser emissions arise. The SiO masers, at distances of a few astronomical units, are nearest the center of a star and are in regions where the number of hydrogen molecules per cubic centimeter is about 10 billion. The OH masers, at distances of thousands of astronomical units, are in the lowest density regions, with number densities of hydrogen molecules of roughly 10 million per cubic centimeter.

The energies of the levels involved in the SiO maser transitions are affected slightly, but measurably, by the magnetic field in a stellar outflow; observations of the wavelengths of the emissions provide information about the strengths of the magnetic fields in the SiO maser spots. The magnetic pressure in such a spot exceeds the thermal pressure, and the Alfvén speed (cf. section 5.2) is higher than the speed at which the maser spot moves away from the star. The relatively large sizes of the magnetic pressures and Alfvén speeds almost certainly provide clues to the dynamics of the maser spot formation. One suggestion regarding the formation process is that the magnetic properties of the spots are consistent with the magnetic field plunging from the spots and other clumps in the outflows down into the stellar atmospheres. Plunging of the magnetic field may couple the dynamics of material with densities varying by a factor of about 1,000 or even much more. If so, the impulse of radiation pressure on dust well above the star's surface is being transferred down to material near the surface.

OH maser lines are also used to measure magnetic field strengths, but more often than not this is done in the context of the environments of young high-mass stars (cf. section 6.2).

7.7 Winds from the Nuclei of Planetary Nebulae

As described in chapter 3, the evolution of low-mass stars through the red giant and AGB stages of stellar evolution leads to the phase in which the red giant or AGB envelope is expelled from the star (cf. section 3.3.1). The stellar core left evolves at roughly constant luminosity and with a rapidly increasing stellar surface temperature. Observations show that these stars are losing mass at between roughly one tenth of a millionth (10^{-7}) and one hundredth of a millionth (10^{-8}) of a solar mass per year, at speeds of up to 4,000 to 5,000 km s^{-1}. Such winds are due to radiative line driving as discussed in section 7.3. Multiple scattering is not required. The very high speeds reflect the high escape speeds of these relatively compact, dense stellar cores.

7.8 Cometary-Like Tails in Planetary Nebulae

As seen in plate 3, the Eskimo Nebula is clumpy, and tails are clearly discernible features of many of the clumps. A number of planetary nebulae have clumps with such long thin tails. Some believe the clumps to have formed in the same fashion as the maser spots described in section 7.6 and to have survived long enough to move to their present positions. Long survival requires the sound speed in the object to be small. This implies low temperatures that obtain only in molecular regions. Thus, the observed gas in a clump would be restricted to an ionized sheath that is replenished by the slow evaporation of the underlying cold material. Two separate explanations have been offered for the sustained existence of the long thin tails. Both are very similar to those described in the context of cometary-like tails in Orion (cf. section 6.7).

SELECTED REFERENCES

Charbonneau, P., and MacGregor, K. B. "Stellar Winds with Non-WKB Alfvén Waves. II. Wind Models for Cool, Evolved Stars." *Astrophysical Journal* 454 (1995): 901–909.

Cherchneff, I. "Dust Formation in Carbon-Rich AGB Stars." In *The Molecular Astrophysics of Stars and Galaxies*, ed. T. W. Hartquist and D. A. Williams, 265–283. Oxford: Clarendon Press, 1998.

Hartquist, T. W., and Dyson, J. E. "The Origin of Strong Magnetic Fields in Circumstellar SiO Masers." *Astronomy and Astrophysics* 319 (1997): 589–592.

Lucy, L. B., and Abott, D. C. "Multiline Transfer and the Dynamics of Wolf-Rayet Winds." *Astrophysical Journal* 405 (1993): 738–746.

Lucy, L. B., and Solomon, L. B. "Mass Loss by Hot Stars." *Astrophysical Journal* 159 (1970): 879–894.

8

SUPERNOVAE AND THEIR REMNANTS

An explosion eventually occurs in a highly evolved single star of sufficient mass or in a white dwarf that has accreted enough material from its companion in a binary system. Explosions give some individual stars optical luminosities that for a couple of months are more than a billion times that of the Sun. These sources, when in our Galaxy, have appeared suddenly as very bright, new objects in the sky. They are called supernovae because of their power and because the Latin word for new is nova, as mentioned in section 1.3. (Supernovae differ from novae, which are introduced in that section and are treated more fully in section 11.3.) Supernovae within the Milky Way have been rare during historical times. The last one was observed in 1604 and is named after Kepler.

The nature of supernova explosions is a prime topic in this chapter. Some supernovae are used to estimate distances to faraway galaxies in order to infer how the Universe expands and its ultimate fate. Despite the power of a supernova, molecules have been observed to form in the ejecta of one of them just several months after the explosion. The molecular emission has been used together with theoretical models to make some inferences about the early propagation of that explosion through the progenitor star's stratified layers of nuclear ash. The ejecta of a supernova interact with the surrounding interstellar matter to form a *supernova remnant*, which passes through several well-defined stages of evolution, during which X-ray emission and

massive that further burning begins. Because the electrons in the white dwarf are degenerate, the increase in temperature does not cause a significant increase in the pressure (cf. section 3.3) or expansion, but does cause the rate of burning to go up, which reinforces the rise in temperature. This feedback between increasing temperature and increasing rate of the nuclear burning reactions, together with the associated lack of expansion of the material, results in a thermonuclear runaway, and an explosion ensues.

The Type Ib supernovae constitute another class. Though they share some observational characteristics with the Type Ia supernovae, the Type Ib explosions probably occur in a process similar to the explosion mechanism of Type IIs, that is, in single stars.

A Type II supernova is the more energetic of the types we have described. Each releases about 10^{44} Joules of kinetic energy, which is comparable to the total energy that will have been emitted by the Sun in its lifetime. The material ejected in each Type II supernova has a mass several times the Sun's mass and travels initially at a speed of up to 10,000 km s^{-1}.

8.2 The Brightness of a Supernova and the Expansion of the Universe

In the 1920s Edwin Hubble discovered observationally that the Universe is expanding. Ever since, one of the questions to intrigue scientists (and nonscientists) most is that of the ultimate fate of the Universe. Will it expand forever, or is the gravity within it strong enough to cause it to cease expanding and to collapse back upon itself to a singularity whose nature will be governed by quantum gravitational effects? Supernovae in very distant galaxies have become the subject of much study in attempts to address this fascinating question.

To investigate the expansion of the Universe, Hubble had to measure the distances to and recession speeds of many galaxies. The main principle behind measuring distances to other galaxies is based on the identification of a class of sources that possess well-known intrinsic brightnesses that vary little between individual members of the class; a source belonging to such a class is often called a *standard candle*. A measurement of a standard candle's apparent brightness at Earth together with its known intrinsic brightness can be used to derive the distance to the galaxy containing it. The recession speed of a galaxy is inferred from the determination of how redshifted any spectral features originating in the galaxy are. (We introduced the concept of redshifted radiation in sections 1.3 and 5.1, as well as the associated relationship between recession speed and how much a perceived wavelength or frequency differs from that intrinsic to the source.) Hubble found a linear relationship between the distance to a galaxy and its recession speed, as shown in figure 8-2.

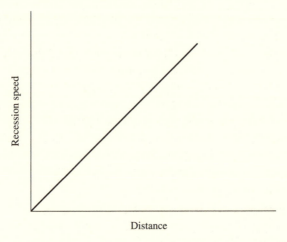

FIGURE 8-2   The Hubble Expansion. A linear relationship between recession speed and distance would be expected if all galaxies were ejected from the same point at the same time and none either accelerated or decelerated.

A linear relationship between recession speed and distance is exactly what one would expect if all galaxies were ejected from the same point at the same time and each moved at a velocity that never changed with time. So on the basis of a linear relationship between recession speed and distance, one would infer that the expansion of the Universe is neither accelerating or decelerating. However, Hubble made observations of galaxies that are moving relative to us at speeds that are only a small fraction of the speed of light. A galaxy near the horizon of the accessible part of the Universe would move away from us at a speed close to that of light. Hence, in cosmological terms the galaxies used by Hubble are relatively nearby compared to the extent of that part of the Universe that is in principle accessible to observation. The distance to which Hubble could investigate the expansion of the Universe was limited to some degree by the instruments of his time, but the choice of standard candle can also limit a study. In the second and third decades of the 20th century, the discovery of the existence of galaxies external to our own, and measurements of distances to them, were facilitated by the use of a class of stars called *Cepheid variables* as standard candles.

For years some scientists have claimed that the distances to very far-away galaxies are well determined through the adoption of Type Ia supernovae as standard candles. A supernova is a particularly bright type of single source, and one that occurs in even a very far away galaxy is visible. Thus, if each and every Type Ia supernova has the same intrinsic brightness as each and every other Type Ia supernova, such supernovae would be excellent standard candles. On the assumption that they are, the measurements of the recession speeds of galaxies compared to their distances have been greatly extended to include galaxies moving at good fractions of the speed of light from us.

The observation of these very far away galaxies allows study of the Universe's expansion long ago, because the light that arrives now from such a galaxy took a long time to reach us. If the expansion of the Universe were decelerating, very distant galaxies would be expected to have recession speed-distance points representing them that lie above the straight line in figure 8-2. Some researchers have argued that based on data obtained from the employment of supernovae as standard candles, the points corresponding to distant galaxies lie below the line in figure 8-2 and that this is an indication that the Universe's expansion is accelerating.

Accelerating expansion would be expected from Einstein's general relativistic theory of gravity if a *cosmological constant* of sufficient magnitude were included. Einstein first introduced a cosmological constant in his theory of general relativity because in its original form without that constant, the theory predicted that the Universe should expand or contract but could not be static; a cosmological constant term of the correct magnitude made the theory consistent with the Universe's being static. Of course, Hubble showed that the Universe is expanding, which led Einstein to consider his introduction of the cosmological constant a huge blunder. However, particle physicists now believe that there are good theoretical grounds for the expectation that the cosmological constant is nonzero and possibly important for the Universe's evolution. Even so, there is some chance that the supernovae are not serving as reliable standard candles and the inferred acceleration in expansion and attendant consequences for the value of the cosmological constant may not be secure.

8.3 The Formation of Molecules in Type II Supernova Ejecta

As stated at the start of this chapter, supernovae in the Milky Way have been infrequent during the last thousand or more years. However, in 1987 one occurred in one of the two Magellanic Clouds, dwarf galaxies orbiting the Galaxy at distances of about several hundred thousand light years, which is roughly 10 times the Earth's distance from the center of the Galaxy. This Type II supernova in the Large Magellanic Cloud, designated SN 1987A, was observed thoroughly, and emission from molecules created in its ejecta were detected from 110 to 574 days after SN 1987A was first observable.

Though still travelling at many thousands of kilometers per second, at 110 days of age the several solar masses of the supernova ejecta had cooled from expansion and radiation (due to electron impacts on atoms and ions populating excited states that decayed by photoemission) to several thousands of degrees Kelvin. At 110 days the oxygen-rich part of the ejecta, overlapping the carbon and neon layers, contained about a billion nuclei in each cubic centimeter, and the density was continuing to drop. An indefinite drop in temperature was not occurring, because the ejecta contain radioactive cobalt and other radioactive species produced in the explosion. The radioactive decay of this material provides a source of heat.

At temperatures as low as a few thousand degrees molecules can form and survive. One of the key molecules observed to exist in the SN 1987A ejecta is CO. Its primary mode of formation in an environment like the SN 1987A ejecta is through a radiative association reaction. In that reaction, a neutral carbon atom and neutral oxygen atom near one another lose energy and momentum through the emission of a photon, allowing them to bind together in a single particle of CO. In the ejecta an important mechanism for the destruction of CO may have been its reaction with ionized helium, He^+, produced from neutral helium as a consequence of the radioactive decays of an unstable variety (isotope) of cobalt. The results of theoretical models put an interesting limitation on the amount of helium that could have been present in the part of the ejecta where the CO formed. The presence of too much helium would have resulted in the existence of less CO than is consistent with the observational data.

The constraint on the amount of helium in the parts of the ejecta containing CO allows inferences about some of the dynamics in the supernova explosion, at phases arising shortly after the possible bounce and copious neutrino emission. A look at figure 8-1 reminds one that before the supernova explosion, layers rich in helium and rich in oxygen as well as carbon were well separated. However, the compressibility and neutrino effects result in material from the core region being driven toward the layer just outside the iron core. As a result, a shock propagates through the overlying layers. The interfaces between these layers become highly distorted as the shock passes, due to an effect called the Rayleigh-Taylor instability. The distortion of the interface between two layers is depicted in figure 8-3.

The development of the Rayleigh-Taylor instability can be observed in a type of toy that many will have seen. Such a toy contains fluids of two colors. One fluid is denser than the other and, consequently, with time will always settle to the side of the toy turned toward the floor. If after the denser fluid is fully settled the toy is quickly turned over so that the denser fluid is on the top side, the interface between the two regions of different color will become distorted. Finger-shaped protrusions of the denser fluid will appear and grow.

From the description of the Rayleigh-Taylor instability's effects in the toy, one sees that a prime condition for it to occur is for a denser material to lie above another material in a gravitational field. What provides the gravity in the system consisting of the shock propagating through the strata of nuclear ash? The gravity due to the central core is not important in the shocked gas and, in any case, is in the wrong direction as the denser material is nearer

(a)

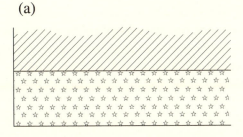

(b)

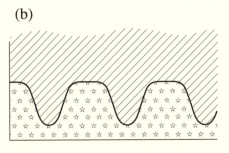

FIGURE 8-3   The Rayleigh-Taylor Instability. Panel (a) shows an initial configuration in which dense material is supported against the gravitational field by the less dense material below it. In this initial configuration the interface between the two substances is flat and perpendicular to the gravitational field. Panel (b) shows a state to which the system evolves. The initial configuration is unstable and the interface between the two substances distorts with prominent "tongue-like" protrusions of the denser material extending in the direction of the gravitational field.

the core whereas for the instability to act the less dense material must lie in the direction of the gravitational field. Thus, if the instability is acting there must be an effective gravitational field pointing outward from the core, not in toward it. The source of the effective gravity is the deceleration of the shocked gas as the shock is slowed by its passing through more material. By imagining being in an elevator, one can understand how the deceleration can serve as an effective gravity. When the elevator accelerates upward a passenger feels heavier. In an elevator, the extra effective gravity is in the direction opposite the acceleration. Deceleration of the shock in the nuclear ash in the outward direction is the same as an acceleration in the inward direction; hence, the effective gravity in the shocked gas is in the outward direction. This is the direction in which less dense material is located, and the instability grows.

Given that such extreme distortions at the interfaces between layers develop as the shock driven by the explosion passes through the strata of nuclear ash, one might wonder whether thorough mixing of material between the layers occurs. The observations of CO molecules in the supernova ejecta show that despite the growth of the Rayleigh-Taylor instability, thorough mixing between the helium-rich stratum and that rich in both oxygen and carbon does not take place.

8.4 The Supernova—Interstellar Matter Interaction

An object created by the interaction of a supernova with the interstellar matter is called a supernova remnant (SNR). The evolution of an SNR has the following phases:

Free expansion
Adiabatic expansion
Pressure-driven snowplough
Momentum-driven snowplough
Contraction

During the earliest phase, the ejecta–interstellar matter system is very much like a windblown bubble (cf. figure 4-6). A double shock structure exists with one shock propagating outward into the ambient medium and the other decelerating the ejecta. This is called the *free expansion* phase, because until it is over at least some of the ejecta have been unaffected by the presence of the interstellar gas. The free expansion phase lasts for a time that is roughly comparable to that required for the ejecta to expand to fill a volume containing a mass of interstellar material comparable to their mass. Consider ejecta with a mass five times that of the Sun travelling at 5,000 km s^{-1} into a clump in a giant molecular cloud (cf. section 5.1) and assume the clump to have a number density of 1,000 hydrogen nuclei per cubic centimeter. Then the free expansion phase ends after roughly 100 years, when the ejecta have travelled somewhat less than a light year. Much of the ejected and swept-up material will have been heated to temperatures of over 100 million K. The most tenuous regions in interstellar space are filled with gas having a density about one million times lower than the density that we assumed above for a clump; if a supernova occurs in such a tenuous region the free expansion phase lasts for about 10,000 years by which time the ejecta have travelled roughly 100 light years.

The heating of the gas in the two shocks during the free expansion phase results in the gas emitting X-rays, with wavelengths typically in the range of about 0.1 to 10 nm (visual light has wavelengths centered on about 550 nm). At temperatures above 10 million K most of the emission is due to the process of *free-free emission* or *bremsstrahlung emission*. The reason that the description *free-free* is used is that the process involves the slowing down, by the emission of a photon, of an electron that is unbound to an ion (and hence free). In this process the electron is still unbound to the ion after the slowing down occurs. The mechanism by which free-free emission occurs is illustrated in the first panel of figure 8-4. The German word *bremsstrahlung* means "braking radiation," so it is also an appropriate term to use in connection with the process. Of course, a thorough understanding of radiative mechanisms important in astrophysics came only after clever experimental approaches were developed to study them and relevant theoretical calculations based on quantum theory were performed. Figure 8-5 shows a map of some of the X-ray emission originating in an SNR, still in its free expansion phase.

After the free expansion phase is over the pressure of the gas heated during that phase drives further expansion. Usually in the Galaxy, an SNR propagates into a low enough density region that it loses only a small fraction of its energy due to the emission of radiation, until it has expanded consider-

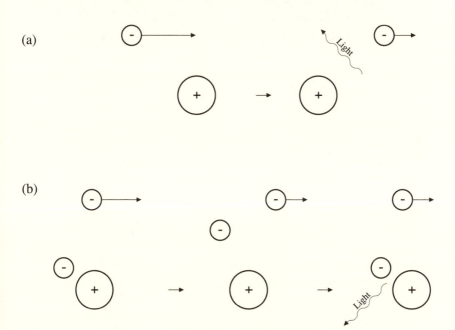

FIGURE 8-4   Radiative Mechanisms. Panel (a) illustrates the process of free-free emission or *brehmsstrahlung*. A free electron approaches a charged nucleus, usually H⁺; in the interaction the electron is slowed by the emission of a photon. Panel (b) represents the process of electron impact–induced radiative emission. A free electron approaches an atom or ion containing one or more electrons. Interaction causes the slowing of the free electron and the excitation (increase in energy) of an electron in the atom or ion. Finally the excited electron returns to the lower state by radiation of a photon.

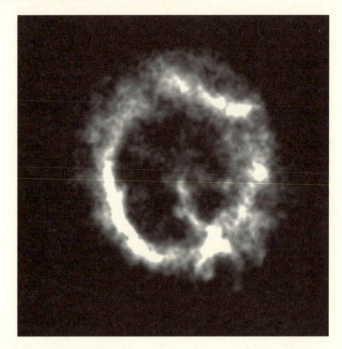

FIGURE 8-5   An X-ray Image of a Supernova Remnant (E0102–72). The remnant is roughly 1,000 years old and has a diameter of about 40 light years. Image courtesy of NASA/CXC/SAO.

ably relative to its size just at the end of the free expansion phase. As long as energy loss due to radiation is negligible and the free expansion phase has ended, an SNR is said to be in the *adiabatic* phase. (In practice, the word adiabatic implies that the total energy of the system remains the same.) If the ambient material is uniform the radius of the SNR will increase in proportion to its age to the two-fifths power, while its expansion speed will drop in inverse proportion to its age to the three-fifths power. An SNR expands at a couple thousand kilometers per second as it enters the adiabatic phase and at a couple hundred kilometers per second when its age is close to 50 times what it was at the start of the adiabatic phase.

During the adiabatic phase an SNR cools primarily due to its expansion. It is bounded by the shock that its expansion drives into the surrounding medium. The SNR is simply a bubble of hot gas surrounded by its shock. The temperature is highest at the SNR's center and decreases toward its outer edge.

When the expansion speed drops to 1,000 km s^{-1}, the gas that has just passed through the shock has a temperature of somewhat over 10 million K. In gas at this and lower temperatures, most of the energy radiated away is carried by photons created by electrons impacting on ions, causing electrons in these ions to jump to excited states; these excited states decay through the emission of radiation in discrete lines. The electrons in the ions are in bound states whether they are in the ground states or in excited states, and quantum theory implies that all bound states have discrete energies rather than a continuum of possible energies; it is the discreteness of the bound state energies that causes the radiation to be in clear well-defined lines. The mechanism of radiating due to electron impact induced excitation of a bound state is illustrated in the second panel of figure 8-4. The ions involved in the cooling obviously have electrons, but at temperatures of over 100,000 K, most of the ions significant for the cooling carry several or many positive charges each; the number of positive charges that an ion carries is equal to the difference in the number of electrons it has and the number the corresponding neutral atom has. For instance, in gas at 1 million degrees Kelvin, important X-ray lines arise from oxygen ions with only one or two electrons left on each, rather than the eight that a neutral oxygen atom has.

When the expansion speed of an SNR falls to 250 km s^{-1}, gas that has just passed through the leading shock is at a temperature only a bit below 1 million degrees Kelvin. The SNR begins to lose substantial amounts of its energy through radiation. It leaves the adiabatic phase and enters the *pressure-driven snowplough* phase. For SNR gas at temperatures above 100,000 K, the time required for loss of a good fraction of its energy through radiation decreases rapidly as the temperature drops; this effect causes gas at the outer edge of the SNR during this phase to cool most rapidly, leading to the formation of a relatively cold thin shell around interior hot gas with temperatures in excess of 10 million K. The shell can be thought of as a snowplough pushed into the ambient medium by the pressure of the hot gas inside it.

By the time that an SNR has expanded to double or treble the volume that it had when entering the pressure-driven snowplough phase, it has experienced enough loss of energy by radiation that its expansion is no longer driven by the pressure of the hot gas remaining in it. Rather its expansion continues because of the inertia that the shell has. The remnant is in the *momentum-driven snowplough* phase. Of course, the expansion of an SNR will finally halt as a consequence of the ambient medium having a pressure that eventually contains it. Once expansion stops, a *contraction* phase begins as the pressure inside the SNR continues to drop due to cooling by the emission of radiation. For a typical SNR in the Galaxy, the contraction phase will not begin until the SNR has reached a radius of many hundreds of light years and an age of well over a million years.

8.5 Some Conditions and Processes That Can Alter Supernova Remnant Evolution

A number of preexisting conditions and processes not mentioned in the previous section can also affect the evolution of an SNR.

Above, we assumed that the ambient medium is uniform. In reality, a Type II supernova is likely to occur inside the remains of a bubble blown by the wind of its progenitor star. If the progenitor star initially contained fewer than 60 solar masses, some regions of the bubble will contain clumps of stellar material lost in 10–30 km s^{-1} winds during the red supergiant phase (cf. section 3.3.2). If the supernova occurs in a dense enough region in an interstellar cloud and such a bubble has been blown, its SNR will undergo a transition from the free expansion phase to the pressure-driven snowplough phase when the windblown bubble's shell of swept-up ambient material is encountered. The adiabatic phase will be bypassed. This sort of SNR evolution is likely to be more relevant in the starburst regions described in section 9.6 than in typical regions in the Milky Way.

In the previous section, we described the free expansion phase as though the ejecta are smoothly distributed, whereas clumps of ejecta have been seen in the Cas A SNR. It is still in the free expansion phase and the clumps contain little hydrogen; instead they are composed primarily of oxygen and other products of advanced nuclear burning. They move at 6,000 km s^{-1}, but a more smoothly distributed component of the ejecta probably expands at about 10,000 km s^{-1}. Clumpy ejecta will drive shocks into an ambient medium, and for phases of evolution after the free expansion phase, the clumpiness of the ejecta has little effect on the general characteristics of the current SNR evolution.

However, clumpiness of the ambient medium can also greatly affect SNR development. In the previous section, we described the evolution of an SNR picking up additional mass by material being swept through its leading surface. If the ambient medium is clumpy, as the interstellar clouds cause the interstellar medium to be, the leading edge of an SNR will travel well past a cloud before the cloud is totally evaporated into the hot, moving SNR gas. Evaporation may be driven by a number of effects, but independent of the details of the evaporation mechanism, a very thin, well-defined shell does not ever develop around an SNR that picks up mass primarily by evaporation rather than by sweeping it up.

The flow of hot material around a cool cloud causes pressure differences on the surface of the cloud, and this probably drives cloud evaporation in many circumstances. Another of the several mechanisms most discussed as a means of driving evaporation is heat conduction. Heat flows from a hot region to a cold region, just as it will spread to the metal handle of a pan on a hot burner of a stove. Heat conduction might cause the heating of outer layers of a cloud and their subsequent expansion into the hot SNR gas.

Heat conduction has also been suggested as a key process in the evolution of an SNR moving into a smooth ambient medium containing no clumps. Conductive transfer of heat to colder outer regions may lead to a significant lowering of the temperature in the central regions of an SNR. This would lower the emission rate of the more energetic X-rays from an SNR and reduce the duration of the pressure-driven snowplough and subsequent phases.

8.6 Cosmic Ray Acceleration in Supernova Remnants

Roughly one out of every one billion nuclei in interstellar space moves at a speed of at least about half the speed of light. Such particles are cosmic rays, which also include high-energy electrons and positrons. For the moment we focus on the cosmic ray protons, many of which are accelerated to relativistic speeds in SNRs. Figure 8-6 shows the dependence measured at Earth of the number density of cosmic rays on the energy per cosmic ray.

The existence of an unexplained source of ionization at the Earth's surface was known around 1900, from the fact that even very well-insulated electroscopes would eventually discharge—some form of radiation was penetrating them. In 1912, V. F. Hess flew an electroscope on a balloon and found that ionization is induced more rapidly at high altitude, a result

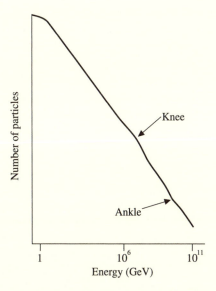

FIGURE 8-6   The Cosmic Ray Spectrum. The number of particles having an energy per particle differing less than one electron volt from a specified energy per particle plotted as a function of that specified energy. Two features called the *knee* and the *ankle* are indicated. Acceleration up to the energies associated with the knee occurs in supernova remnants. Adapted from a figure provided by S. Swordy for the Auger Collaboration, *The Auger Project Design Report*, 2nd ed. (Chicago: Fermi National Accelerator Laboratory, 1997).

consistent with the radiation having an extraterrestrial origin and being attenuated by propagation downward through the Earth's atmosphere. Major theoretical advances on the explanation of the mechanism responsible for the energization of cosmic ray protons were made nearly simultaneously, and independently, by several groups in the mid-1970s. All of these groups identified shocks as the sites of the main acceleration, and many recognized quickly that shocks in SNRs are likely to be of considerable importance for the acceleration of cosmic rays up to energies of around 100,000 GeV, or possibly somewhat higher.

A GeV is one billion electron volts. An electron volt is the energy that an electron will have after it has passed through a potential difference of one volt if it is accelerated from rest. A typical chemical bond has an energy of several electron volts associated with it; the binding energy of an electron in a hydrogen atom is 13.6 electron volts. The rest energy of a proton, which is given by Einstein's famous relationship $E = mc^2$, is about 1 GeV, and the kinetic energy of a proton is about 1 GeV if it is moving at about 80 percent of the speed of light. A proton with an energy of 100,000 GeV moves at a speed that is less than one part in 10 billion slower than the speed of light. When we give the energy of a cosmic ray, we are referring to its kinetic energy only and neglecting the energy associated with its mass through Einstein's result.

The acceleration of cosmic ray protons in shocks is due to their interaction with hydromagnetic waves on either side of the shock front. If it travels toward the upstream direction, a cosmic ray interacts with waves that are moving in the opposite direction. In the interaction the cosmic ray "bounces" so that it moves in the opposite direction; in the bouncing process, it gains energy as a less massive particle does when struck by a much more massive object moving in the opposite direction. Now the cosmic ray moves toward the downstream gas; its speed is much greater than that of the shock so the downstream gas appears to the cosmic ray to be moving toward it. Another bounce occurs and more energy is gained by the cosmic ray. As long as the bouncing continues to occur the cosmic ray will gain more and more energy.

The acceleration by bouncing between magnetic irregularities moving toward one another is called the *first order Fermi mechanism* after the brilliant Italian physicist Enrico Fermi. Amongst his many great contributions was his leadership, during the Second World War, of the team that achieved the first sustained nuclear chain reaction controlled by mankind.

There are limits to how much bouncing occurs. A particle of sufficiently high energy will not be significantly deflected by a magnetic irregularity of too small of an extent. For an SNR the largest linear extent that an irregularity can have is about the SNR's radius. So the radius of the SNR and the magnetic field strength in the gas swept up by the SNR set a limit to the energy to which a proton can be accelerated by the bouncing between irregularities. These and related constraints imply that an individual SNR may not accelerate cosmic ray motions to energies very much above 100,000 GeV, the lower energy of the "knee" feature in the cosmic ray energy distribution shown in figure 8-6. We return to the origin of higher energy cosmic rays in sections 9.5 and 10.8.

As mentioned above, cosmic ray electrons also exist. They, like protons, are accelerated in SNRs and their spiralling in the magnetic fields of these objects gives rise to radio emission observed at wavelengths of about a few centimeters to many tens of centimeters. Such radiation due to the motions of energetic electrons in magnetic fields is called synchrotron radiation and is mentioned again in sections 9.6, 10.8, and 11.4.

Despite their small number compared to that of the total number of inter-stellar nuclei and electrons, the cosmic ray protons with energies of roughly 1 GeV contain about as much kinetic energy as the thermal energy of all the rest of the interstellar gas. A cosmic ray *pressure* is associated with their energy, and it is important for the evolution of an SNR. The gross description of SNR evolution is not altered by the consideration of cosmic rays, but the volume to which an SNR grows and its peak temperature are reduced by the presence of cosmic rays. The dynamical effects of cosmic rays figure in section 9.5.

SELECTED REFERENCES

Cioffi, D. F., McKee, C. F., and Bertschinger, E. "Dynamics of Radiative Supernova Remnants." *Astrophysical Journal* 334 (1988): 252–265.

Dyson, J. E., and Hartquist, T. W. "Mass Loaded Astronomical Flows—III. The Structure of Supernova Remnants and the Local Soft X-ray Background." *Monthly Notices of the Royal Astronomical Society* 228 (1987): 453–461.

Dyson, J. E., and Williams, D. A. *The Physics of the Interstellar Medium.* 2nd ed. Bristol: Institute of Physics Publishing, 1997.

Ellison, D. C., and Reynolds, S. P. "Electron Acceleration in a Nonlinear Shock Model with Applications to Supernova Remnants." *Astrophysical Journal* 382 (1991): 242–254.

Hartquist, T. W., and Williams, D. A. *The Chemically Controlled Cosmos.* Cambridge: Cambridge University Press, 1995.

Jones, F. C., and Ellison, D. C. "The Plasma Physics of Shock Acceleration." *Space Science Reviews* 58 (1991): 259–346.

Liu, W. "Supernova Chemistry." In *The Molecular Astrophysics of Stars and Galaxies,* ed. T. W. Hartquist and D. A. Williams, 413–434. Oxford: Clarendon Press, 1998.

Raymond, J. C. "Observations of Supernova Remnants." *Annual Review of Astronomy and Astrophysics* 22 (1984): 75–95.

Riess, A. G., et al. "Tests of the Accelerating Universe with Near-Infrared Observations of a High Redshift Type Ia Supernova." *Astrophysical Journal* 536 (2000): 62–67.

9

GALACTIC WINDS, STARBURST SUPERWINDS, AND THE EPOCH OF GALAXY FORMATION

In a spiral galaxy, like the Milky Way, most optically bright material is in a disk. However, such a galaxy does contain matter in regions high above the disk. Those regions constitute the halo of a galaxy. Gas heated in supernova remnants can escape into the halo of a galaxy and possibly flow out as a galactic wind. Observations of the absorption of ultraviolet radiation emitted by high-mass stars in the halo of the Milky Way reveal the presence of moderately to highly charged ions far above the Galaxy's disk. X-ray emission from the halos of galaxies similar to ours shows that they also contain such species.

Extended distributions of neutral clouds as well as charged ions around galaxies may account for the origin of many of the "narrow" absorption features detected in the spectra of quasistellar objects and other active galaxies, some of which are the most distant objects to have been discovered (cf. section 10.2). Some of the properties of these narrow absorption features most likely provide clues about the formation and evolution of galaxies. They may even give some insight into the global properties of the Universe.

The Milky Way may have a fountain-like flow, in which hot material flows upward and then cools to form clouds that fall back to the disk, rather than a galactic wind. Fountain models have been suggested to explain the origin of the high velocity clouds (HVCs, cf. section 9.4), though some believe that these objects formed far from the Milky Way and are the building blocks of galaxies rather than the products of them.

The pressure of cosmic rays can drive flows into the halo of a galaxy. Perhaps even more fascinating is the possibility that winds of galaxies may play a role in the acceleration of cosmic rays to ultrahigh energies, up to about a billion times that associated with the rest-mass of a proton by Einstein's energy-to-mass relationship.

Some galaxies have bursts of star formation concentrated within about a thousand light years of their centers. Such *starburst* galaxies have superwinds that radiate brilliantly in X-rays. At the epoch of galaxy formation, most galaxies were probably starburst galaxies or contained even more extreme starburst activity than current starburst galaxies. Their superwinds affected the subsequent formation of structure in the young Universe.

9.1 How Hot Gas Might Escape from the Galactic Disk

The Milky Way is a spiral galaxy with a flat disk (cf. section 5.1), which contains most of the high-mass stars that evolve to produce Type II supernovae. Type Ia supernovae are not quite so confined to the disk, but the bulk of supernova energy is injected into it. Gas heated by supernova remnant (SNR) shocks may rise to great heights above the disk.

In section 3.1 and particularly with figure 3-1, we pointed out that the thermal pressure of gas acts to support the gas in a gravitational field. For a given gravitational field, the extent of a gas will increase with its temperature (cf. section 7.1). In the gravitational field of the Milky Way, gas at 10,000 K will extend many hundreds of light years on either side of the midplane of the disk. Gas at 100,000 K will extend many thousands of light years, whereas gas at temperatures much in excess of several million degrees Kelvin can escape from the Galaxy if it does not cool while rising.

Some of the volume of interstellar space in the disk of the Milky Way is filled with gas at 10,000 K. The fraction of the volume that is filled with such gas has important consequences for the escape of hotter gas into the halo. If that fraction is large and supernovae are isolated events, then a SNR may be viewed as an island in a sea of 10,000 K gas, and hot gas within the SNR will not escape. However, if supernovae are sufficiently frequent, the hot gas within them will fill a larger fraction of the space than the 10,000 K gas does, and some of the hot gas will be able to rise above the disk of the Galaxy. For about a quarter of a century there has been intense discussion on the fractions of space in the disk filled by the 10,000 K gas and hotter (roughly 1 million K) gas. Much of that discussion has been based on results of theoretical models of SNR evolution and on radio wavelength observations of the distribution of 10,000 K gas. Estimates for the fraction of that space filled by gas at 10,000 K have ranged from about one tenth to eight tenths; a comparable range of estimates have been suggested for the fraction filled by hot gas. Currently, those arguing for a fraction less than one half being filled by hot gas have probably succeeded in convincing more of their colleagues; if these advocates are correct, hot gas in the remnant of an isolated supernova will not escape into the halo.

However, Type II supernovae are usually not isolated. As seen in chapter 6, high-mass stars, which are the progenitors of Type II supernovae, form in clusters. Consequently, several tens of supernovae can occur within a hundred light years of one another within a period of 10 million years, the typical timescale for a supernova progenitor to evolve from birth to become a supernova. Thus, supernovae can explode in a limited region one after another before radiative losses become important. The remnant of such a burst of supernovae is called a *superbubble*.

The Earth is probably surrounded by a superbubble in which gas has been heated repeatedly by subsequent supernovae over a period of several tens of millions of years or more. This circumterrestrial superbubble probably produces much of the X-ray emission observed at wavelengths of about 2.5 to 10 nm. (Recall that visual radiation has wavelengths around 550 nm.)

If a number of supernovae occur in sufficiently rapid succession near enough to one another, their combined remnant expands much further than the remnant of a single supernova. Whereas the maximum radius of the remnant of a single supernova may be somewhat less than the half thickness of the distribution of the disk gas, the remnant of tens of nearly cospatial and coincident supernovae breaks through the disk gas. Once it does so and enters the halo the remnant's expansion velocity outward increases, even though its expansion sideways in the disk continuously slows. Figure 9-1 shows schematically how a superbubble evolves. Superbubbles of multiple supernovae are almost certainly the source of most of the mass leaving the Milky Way's disk and expanding into its halo. In other galaxies that have markedly higher formation rates of high-mass stars, the superbubbles undoubtedly dominate the mass transfer to the halos.

9.2 Observations of Moderately and Highly Ionized Halo Gas

The two main methods for observationally studying interstellar gas at temperatures at or above about 100,000 K are the detection of absorption features and the measurement of extreme ultraviolet and X-ray emissions.

The concept of absorption in spectral lines was first mentioned in section 1.3. Observations of a hotter source of radiation behind a colder gaseous region will show evidence of absorption of radiation by the colder gas. Much of the absorption is in discrete spectral features associated with transitions between bound quantum states of atoms, ions, or molecules in the colder gas.

Absorption features arising in gas in the Milky Way's halo have been observed in the directions of high-mass stars in the halo and of a few par-

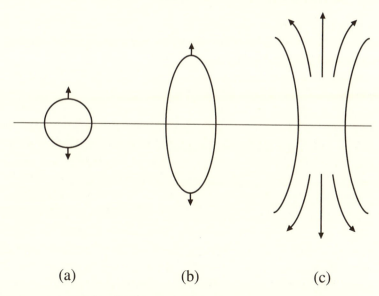

(a)   (b)   (c)

FIGURE 9-1   A Superbubble Expelling Gas into the Halo. Successive stages of development are illustrated from left to right. Panel (a) shows the bubble's boundary when its size is less than the length scale or *scale height* over which the density of the disk of the galaxy is nearly constant. Panel (b) shows the boundary when the half size of the bubble is a couple times the scale height. Eventually the expansion begins to accelerate out of the disk while slowing appreciably in the disk, as depicted in panel (c).

ticularly bright extragalactic sources. The presence of these high-mass stars in the Galaxy's halo is an unsolved problem. The velocity away from or toward the Sun of a high-mass star in the halo is inferred from the redshift or blueshift of its emission features. High-mass stars have much shorter lifetimes than the Sun. The measured velocities are too low for the high-mass stars to have been born in the disk and reached their positions in the halo in their relatively brief lives. The fact that the stars seem to have been born in the halo is baffling because little of the Galaxy's molecular material, in which stars are born, is in the halo. The bright extragalactic sources against which absorption observations are made include the brightest, hottest stars in the Magellanic Clouds, which are two smaller, more irregularly structured, companion galaxies to the Milky Way. The extragalactic sources used in studying Milky Way gas also include several of the brighter active galaxies.

The absorption data that reveal information about halo gas at temperatures in excess of 10,000 K were obtained in observations at ultraviolet wavelengths between about 91.2 nm and 200 nm. The International Ultraviolet Explorer (IUE) (cf. the penultimate paragraph of chapter 1) was used to gather the first such data, and absorption by the species C^{3+}, Si^{3+}, and N^{4+} was observed. The interpretation of such observations depends on theoretical calculations of the fraction of each element that will be in each of its particular ionized states under various physical conditions. That fraction depends upon such variables as the temperature and the strength and spectrum of the radiation field incident upon the absorbing gas. Some standard assumptions often made are that the incident radiation field is negligible and the gas has been at the same temperature for a very long time. Figure 9-2 shows the results of such a calculation for C^{3+} made for these assumptions.

The primary input for a calculation of this type is the set of rates of ionization induced by collisions of electrons with atoms and ions and rates of electron capture by the ions. Much of our knowledge of such rates comes from intensive computational studies based on quantum theory. It was no accident that both Robert Wilson, who designed IUE, and Michael Seaton, who was a key contributor to many of the theoretical and computational developments necessary for such quantum studies, were in the Department of Physics and Astronomy of University College London in 1978 when that satellite was launched. Astrophysics depends on a great deal of fundamental work in physics and chemistry.

The IUE measurements of the C^{3+} and Si^{3+} abundance ratio were first thought to imply a temperature of about 80,000 K. For that temperature, the sort of model of a nonevolving and nonirradiated gas mentioned above gives an abundance ratio that matches the measured ratio. Initially N^{4+}, which is abundant at somewhat higher temperatures, was not detected in very many directions. Thus, originally IUE data were thought to imply a temperature of the halo gas that was too low to drive gas to heights much in excess of

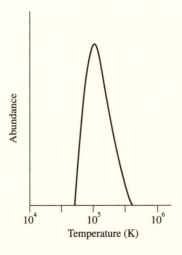

FIGURE 9-2 The Abundance of C^{3+} as a Function of Temperature.

10,000 light years (roughly a third of the Sun's distance from the Galactic Center) into the halo.

The scientific process is one that requires revision of ideas. Subsequently, after more IUE observations were made, the general presence of N^{4+} in the halo was established. It indicates that a range of temperatures obtain in the hot halo gas, whereas the early C^{3+} and Si^{3+} results implied a fairly narrow temperature distribution. Until recently the most generally accepted picture offered to explain the ultraviolet absorption data is that the N^{4+}, C^{3+}, and Si^{3+} is in gas cooling down from higher temperatures through the emission of soft X-ray and extreme ultraviolet radiation. The radiation emitted from the hotter gas actually affects the abundances of C^{3+} and Si^{3+} in the somewhat lower temperature gas. The combination of the existing data and the theoretical models do not provide a strong constraint on how hot the hottest halo gas is. All that can be said with certainty is that it must be at a temperature of above 200,000 K. Detections of O^{5+} in the halo gas have been made recently. However, its distribution is more limited than those of C^{3+}, Si^{3+}, and N^{4+}, perhaps implying that the regions where gas is much hotter than that fill appreciably less than half of the volume. We do have some insight into the density of the hot halo gas. It is probably around one particle per hundred or thousand cubic centimeters.

As mentioned above, the hot halo gas radiates. In the previous chapter, we saw that X-rays from gas in some supernova remnants at temperatures like those of the halo gas are observable. So we might hope to observe some of the X-ray emission from the halo gas. One difficulty in studying such emission from the halo is that we are located in a superbubble that emits the same sort of radiation. When we look outward at emission coming from all directions in the sky, we have problems telling whether the emission is from nearby or distant gas. This is the case in the attempt to detect X-ray emission from Milky Way halo gas. There is one definite indication that some of the observed X-ray emission is from heights above 1,000 light years. The distances to a few clouds of gas almost directly above Earth in the Galaxy are known to be about 1,000 light years from observations of stars in the same directions. (We can determine the distances of stars that are not partly obscured by a cloud and stars that are partly obscured by it.) The lowest energy X-ray emission does not penetrate a cloud. Hence, a comparison of the strength of the lowest energy X-ray emission from the cloud direction to the strength of similar emission from a line of sight near it but not passing through the cloud gives a measure of the fraction of emission coming from material at heights greater than that of the cloud. For a cloud at a height of about 1,000 light years that fraction is roughly one third for X-rays with wavelengths of around 90 nm. However, we do not know what fraction of that fraction is from material that is twice as far away as the cloud as opposed to only 50 percent further than it. Consequently, by the standards of the Galaxy's sizescale, the observations tell us little about the emission of gas that is definitely at great heights in the halo.

Turning from the frustrated attempts to study the Milky Way's halo X-ray emission, astronomers searched for such emission from the halos of nearby galaxies. Such emission has been seen at heights of tens of thousands of light years above the disks of a few nearby edge-on spiral galaxies with many properties similar to those of the Milky Way. By describing it as *edge-on* we mean that a galaxy is orientated such that the line of sight to it lies in the plane of its disk. Observations of a galaxy viewed *face-on* (i.e., such that the line of sight is perpendicular to the disk) would not allow us to tell whether the emission comes from halo gas or from gas in the disk. The observations of X-ray emission from the halos of edge-on galaxies with properties similar to the Milky Way have established that gas at temperatures of at least about one million degrees Kelvin exists in those halos.

9.3 Narrow-Line Absorption Features
Toward Active Galactic Nuclei

In the previous section we mentioned that some of the brightest active galaxies have been used as background sources in studies of the Milky Way halo gas. In the next chapter we treat active galaxies and their nuclei more fully.

Some active galaxies are at distances that are significant fractions of the size of the visible Universe and contain nuclei bright enough to be used as background sources in absorption studies of material located *anywhere* between them and the Earth. Thus, absorption observed against a distant background active galactic nucleus (AGN) sometimes is due to material at cosmological distances (of billions of light years).

The distance to any very far away absorbing material is derived from its line of sight velocity of recession. That velocity is computed from the shift in frequencies of spectral lines from the frequencies they would have if the material were at rest with respect to us. With the use of a plot similar to that in figure 8-2 the velocity can be converted into a distance.

Some of the observed absorption arises in matter moving relative to us at speeds of many tenths of the speed of light and it is such matter that is at a good fraction of the visible Universe's radius away from us. In many cases, the high recession velocities shift features from ultraviolet to visual wavelengths. Consequently, the great collecting areas of ground-based optical telescopes allow the use of much weaker background AGNs to study ultraviolet absorption at cosmological distances than can be observed with the small space-borne ultraviolet telescopes to study absorption by the Milky Way halo gas.

Various types of spectral features are seen toward AGNs. The features arising in the so-called narrow-line absorption regions (NLARs) are sufficiently sharp in frequency that the internal motions in many of the NLARs can have speeds of only tens of kilometers per second, roughly the thermal speed of gas at temperatures of 10,000 to 100,000 K. Higher internal speeds would broaden the features more.

Absorption arising due to the presence of C^{3+}, Si^{3+}, N^{4+}, and O^{5+} in NLARs has been commonly detected. The existence of these species in those regions and their presence in the Milky Way's halo gas might suggest that some of the NLARs are the gaseous components of the halos of galaxies along the lines of sight to the AGNs. This hypothesis, though attractive, does face a serious problem: from the known number of galaxies similar to the Milky Way and the known extent of the Milky Way's hot halo gas distribution, we would expect hot halo gas in other Milky Way type galaxies to create many fewer NLARs than are found. Various solutions to this problem have been offered. For instance, during some earlier epochs of the Universe, more galaxies had considerably higher rates of star formation and supernovae than the Milky Way. These galaxies may have had much larger distributions of halo gas than the Milky Way at the current time. Another suggestion is that the halos of the "dwarf" galaxies or of the objects comprising some particularly visually faint component of the Universe's structure may be responsible for the absorption features. Some of the NLARs are probably intrinsic to the AGNs.

Most of the NLARs were detected originally only in absorption due to hydrogen and are poor in elements more massive than helium. Features of this type are said to belong to the *Lyman-alpha forest*. The absorption is due to the Lyman-alpha transition; it involves a photon with a wavelength of 121.6 nm (in the frame of the absorbing gas), which is 4/3 times the maximum wavelength of radiation that can ionize hydrogen. The concept of a forest is invoked because the number of observed features in Lyman-alpha absorption is so large. If one sees a forest, one is first aware that the number of trees is large before one focuses on individual trees! The number and properties of the Lyman-alpha forest features have provided considerable constraints on structure formation and evolution in cosmological models. Similar constraints are contained in the data for other NLARs but are entwined with information concerning the evolution of galaxies in which stars have already produced larger quantities of elements more massive than helium.

9.4 Lower Temperature Gas at High Latitudes

As stated in section 9.1 some superbubbles break through the gas associated with the disk of the Galaxy and carry hot material into the halo. In section 9.2 we noted that observed abundances of moderately to highly ionized species may be produced in gas cooling by radiative emission from higher temperatures. The cooling continues to temperatures of around 100 K, at which absorption of radiation from stars balances the energy losses due to emission.

Once gas has cooled, thermal pressure is no longer effective in driving the gas upward. Any cooled gas that is not moving at a speed exceeding the escape speed of the Galaxy, of several hundred kilometers per second, will fall back to the disk. The flow of hot gas away from the disk of a galaxy and cooled gas falling back toward it has been called a *galactic fountain*. One is illustrated schematically in figure 9-3.

If a fountain flow obtains in the Milky Way, we would expect to detect cold gas falling toward the disk at high speeds. Some astronomers argue that such infall has been observed. In the early 1960s, astronomers compiled surveys of the distribution of atomic hydrogen in the Galaxy made with radio telescopes tuned to emission from the hydrogen at a wavelength of 21 cm. The survey led to the discovery of apparently infalling material moving at speeds of up to more than 200 km s^{-1}, which is comparable to the speed at which the Sun moves in roughly circular orbit in the Galaxy. The objects were named high velocity clouds (HVCs). A small fraction of them appear to be moving away rather than toward us. Attempts have been made to achieve detailed matches between the observed velocity distributions of the HVCs and those calculated for galactic fountain models.

Though those attempts have been successful in some promising ways, considerable debate over the nature of the HVCs exists. Leo Blitz of the

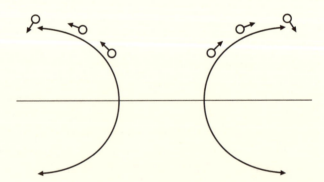

FIGURE 9-3   Fountain Flow. Gas is driven by pressure from the disk. Gravity acting on the gas causes it to decelerate. In the top half of the galaxy, the outflowing gas cools before it reaches its greatest distance from the galaxy's central plane; clouds form in the outwardly moving gas and are eventually slowed by the gravity and fall back toward plane. In the lower half of the galaxy, cooling does not occur before gravity halts the outward flow. Such a dual natured flow could result if a bubble were powered by a supernova or several supernovae that occurred below the central plane.

University of California, Berkeley, and his collaborators have offered some intriguing arguments that the HVCs are primarily a population of extragalactic sources bound to the local group of galaxies. The local group is a cluster of galaxies. A very rich, massive cluster of galaxies is shown in figure 9-4. The local group of galaxies contains one other galaxy similar to the Milky Way and close to 100 dwarf galaxies and has an extent of several million light years. Professor Blitz has suggested that most of the HVCs are falling in the direction of the center of mass of the local group and that objects very much like the HVCs may have been amongst the building blocks of the Milky Way.

If most of the HVCs are extragalactic then there is little evidence for fountain-like flow in the Milky Way. Perhaps, each typical superbubble is powered by a sufficient number of supernovae that the temperature and speed of gas injected into the halo are large enough that the gas escapes and remains hot long enough that little of it becomes neutral. Ultraviolet radiation escaping from the Galaxy and from other galaxies may be important for preventing the Milky Way halo gas from becoming neutral, even if it cools as it escapes.

FIGURE 9-4   Optical Image of a Cluster of Galaxies (MS 1054–0321). It contains over a thousand galaxies and is eight billion light years away. Image courtesy of NASA/STScI.

9.5 Cosmic Rays and the Halo

Heat is not the only means of driving either a galactic fountain or an escaping galactic wind. In section 8.6 we considered cosmic rays. Protons moving at over 50 percent or so of the speed of light constitute only roughly one part in a billion of the interstellar matter but contain well over 10 percent of its kinetic energy. These cosmic rays lose energy through radiation only at a very small rate compared to the other interstellar material. The absence of significant radiative losses means that they can act to propel gas out of the Milky Way even if the thermal pressure of the gas is too diminished, as a consequence of radiative losses, to do so. The Galaxy may well possess a wind accelerated to the escape speed by the transfer of energy from the cosmic rays to other matter.

The Galactic wind may be a source of very high energy cosmic rays. In section 8.6 we pointed out that cosmic ray protons are probably not accelerated to energies much in excess of roughly 100,000 GeV or 10^{14} eV. However, figure 8-6, showing the cosmic ray spectrum, indicates clearly that cosmic rays with energies of well over a million times that exist. Sources other than single supernova remnants must be responsible for the acceleration of cosmic rays above the energies associated with the knee at 10^{14} to 10^{15} eV.

The number of cosmic rays with energies above about 10^{15} eV is sufficiently low that they are not detected directly. Rather their subtle effects on the Earth's atmosphere are detected, just as aurorae and not the solar wind, which induces them, are seen at the Earth's surface. When an energetic cosmic ray hits the nucleus of an atom in a molecule of air, more particles are created. Some of these particles are energetic nucleons, which are protons or neutrons. They, too, can hit nuclei in the air and produce more nucleons. An avalanche of nucleons is created. The nucleons interacting with the air also create particles called muons. The nucleon-muon shower is detected at the ground. The distribution on the ground of the nucleon-muon shower gives information about the particle that triggered the entire shower.

Though the existence of the showers was suspected somewhat earlier by Bruno Rossi, they were investigated thoroughly for the first time in 1938 by Pierre Auger and his collaborators. Since then a number of arrays of detectors have been spread over many square kilometers to study air showers. They have allowed the discovery of cosmic rays with energies of up to 3×10^{20} eV, the energy of a tennis ball travelling at about 100 miles per hour. However, too few of the ultrahigh energy cosmic rays (those with energies of 10^{18} eV and above) have been detected for much information about their origins to be discerned. Thus, a large international team, for which James Cronin of the University of Chicago and Alan Watson of the University of Leeds have served as co-spokesmen, is building a huge array of air shower detectors to be part of a facility called the Pierre Auger Observatory. The 1,600 detectors will be distributed over a 3,000 square kilometer area high in the Argentinian mountains.

Figure 9-5 is a photograph of a prototype detector. It is essentially a large tank of clean water surrounded by some electronics. A shower particle travelling at nearly the speed of light in air will induce the emission of radiation when it encounters the water as it slows to below the speed of light in the water, which is lower than that in air. This radiation is called *Čerenkov radiation* and is registered by the electronics.

The Pierre Auger Observatory will also have instruments to detect optical light and ultraviolet radiation emitted in the sky as a consequence of the interaction of the shower particles with air. The interaction produces π^{0} particles. They are subatomic particles. Each has a mass of about one seventh that of a proton and decays into two photons. Each of these photons is much more energetic than a photon of visual light, and each creates fast electrons in its interaction with the air. These first electrons collide with the air and collisionally induce the emission of visual light and ultraviolet radiation.

FIGURE 9-5   An Auger Detector. The photo was kindly provided by Mansukh Patel, of the University of Leeds, who is pictured toward the center.

Together the particle and light detectors of the Pierre Auger Observatory will almost certainly lead to the discovery of the sources of the ultrahigh energy cosmic rays.

Acceleration to very high energies may occur at the interface of a Galactic wind and the intergalactic medium. The existence of gas between galaxies has been demonstrated most dramatically through the observation of X-rays from clusters of galaxies (cf. figure 9-4). Some clusters of galaxies contain thousands of times as much mass as the Galaxy; typically a large cluster is millions of light years across. The space between the galaxies appears empty. However, detected X-rays having wavelengths of about 0.01 to 1 nm are emitted in that space by gas at temperatures of around 100 million degrees Kelvin. The gas is believed to be heated as it falls toward the center of the cluster. In any one of the largest clusters, infall is supersonic at several thousands of kilometers per second, but is decelerated in a shock as it approaches the gas that has already accumulated nearer the center of the cluster.

In the previous section we mentioned the local group of galaxies. It is not a particularly massive cluster of galaxies. It contains intergalactic matter like other clusters of galaxies, but that matter is at a lower pressure than the X-ray emitting gas in the most massive clusters. If the Milky Way has a wind, its interaction with the intergalactic medium forms a bubble very much like the bubble blown by the solar wind into the interstellar medium (cf. section 4.3). As described in section 8.6, the radius of the bubble and the magnetic field strength of the shock set limits to the energy to which a cosmic ray may be accelerated. The radius of the shock driven by a wind from the entire Milky Way would be many hundreds of thousands of light years. This huge size implies that such a shock probably can accelerate a cosmic ray proton to an energy of as much as about 10,000 times higher than the lowest energies associated with the knee in figure 8-6. That corresponds to an energy of about one billion GeV or 10^{18} eV.

Inspection of figure 8-6 shows that the cosmic ray spectrum continues beyond 10^{18} eV but that there is a feature, called the ankle, in the spectrum. Acceleration in some other type of source probably begins at the energies associated with the ankle. The identity of this type of source is far from certain and we return in the next chapter to a suggestion concerning it.

9.6 The Superwinds of Starburst Galaxies

Some galaxies, very much like the Milky Way in some respects, are very much brighter in their central regions than the Milky Way. Some are called starburst galaxies because of the high rates of star formation near their centers. The birthrate of the massive stars that give rise to supernovae, within several hundred light years (very roughly 1 percent the Sun's distance from the Galactic Center) of the center of a starburst galaxy, can be several to 10 times the rate for the entire Milky Way. The supernova rate in a volume of space that is small compared to that of an entire galaxy can exceed one every 10 years. The hot gas of the overlapping remnants of these supernovae evaporates material from the clouds in which the young massive stars are forming. A superwind with a temperature of about 10 million degrees Kelvin flows at many hundreds of kilometers per second and within about 1,000 light years of the galaxy's center has a number density of about one particle per cubic centimeter.

M82 is a nearby starburst galaxy. Figure 9-6 shows a map of the X-ray emission of its the superwind. Figure 9-7 shows a map of radio emission from

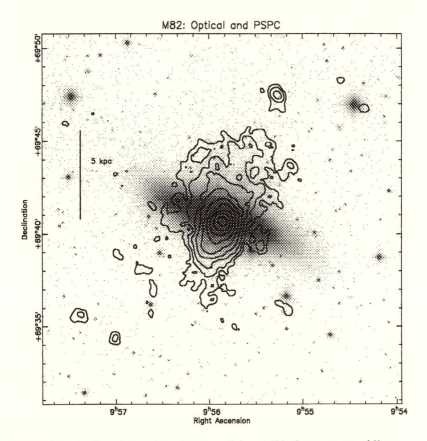

FIGURE 9-6   An X-ray Map of the Starburst Galaxy M82. Contours are of X-ray emission overlaying a gray-scale representation of the optical emission. A line marked 5 kpc indicates a distance of about 15,000 light years. Clearly, the X-ray emission extends over a scale comparable to that of the visible stars in the galaxy. Ian Stevens kindly provided the figure. Details concerning the observation were reported by Strickland, D. K., Ponman, T. J., and Stevens, I. R., "ROSAT Observations of the Galactic Wind in M82," *Astronomy and Astrophysics* 320 (1997): 387–394.

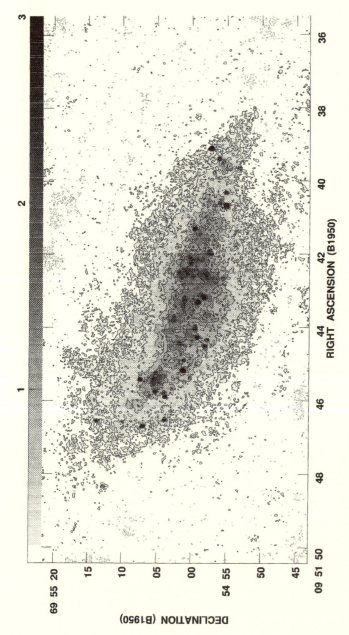

FIGURE 9-7   A Radio Map of the Starburst Galaxy M82. Highest contours are associated with recent supernovae and young supernova remnants. The distance from the top of the field of view to the bottom is about 2,000 light years. The coordinate system used in this figure differs somewhat from that employed for figure 9-6, which is based on the Earth's orientation relative to the stars in the year 2000. Karen Wills generously supplied the figure. The observations are thoroughly described by Wills, K. A., Redman, M. P., Muxlow, T. W. B., and Pedlar, A., "Chimneys in the Starburst Galaxy M82," *Monthly Notices of the Royal Astronomical Society* 309 (1999): 395–403. The map was made available by the courtesy of Blackwell Science.

the starburst region of that galaxy. In the latter figure, remnants of young supernovae are clearly identifiable. The radio emission is due to the spiralling of cosmic ray electrons in the magnetic fields of the supernova remnants (cf. section 8.6's penultimate paragraph).

The means by which a starburst is triggered at the center of a galaxy is not entirely understood. Many of the starburst galaxies (and the active galaxies treated in the following chapter) have nearby neighboring galaxies or are involved in collisions with other galaxies. The interaction between galaxies will distort them. Whereas a typical isolated spiral galaxy has a symmetric structure, allowing orbits in it to be nearly circular about its center, a distorted galaxy's structure leads it to have a gravitational field in which circular orbits are not allowed. The less symmetric orbiting of gas in a galaxy is likely to lead to more collisions between clouds, resulting in the loss of orbital kinetic energy as the clouds do not bounce elastically off one another. The decrease of orbital kinetic energy leads the clouds to fall toward the center of the galaxy in which they are found. Thus, a larger percentage of the interstellar gas is likely to be near the center of a colliding galaxy than in an isolated galaxy.

However, just having lots of gas around may not by itself ensure star formation. For a gas of a certain temperature and a specified mass to collapse due to the gravity it produces, the pressure on its outer boundary must exceed a minimum value (cf. the discussion of the Jeans mass in section 5.3). Consequently a high enough pressure in the medium surrounding the clouds is required for the collapse and star formation to begin, no matter how many clouds of gas have fallen near a galaxy's center.

Once star formation takes place and supernovae begin to occur, the pressure of the intercloud material will increase in response to the heating by supernovae. If the temperature in the clouds is constant, this increase in pressure will cause more and more of the clouds to collapse, leading to more rapid rates of star formation and supernovae; a feedback loop between pressure increase and rising star formation and supernova rates is established, and a runaway or burst will occur!

Certain conditions must be satisfied for this feedback loop to operate. One is implied above: the temperature in the clouds must not increase too much despite the increase of supernova and other activity in the region. The minimum (or Jeans) mass for cloud collapse goes up if the cloud temperature goes up, unless the external pressure increases rapidly enough to counterbalance the effects of the increased cloud temperature. Another condition that must be met for the feedback loop to operate requires that the ratio of the mass in smaller clouds to that in bigger clouds is in a favorable range. If a region contains one cloud that is much more massive than all of the rest near it, that cloud might collapse and produce supernovae, thus driving up the pressure of the intercloud medium without it ever reaching a high enough level to induce the collapse of the smaller clouds.

In any case, starbursts do occur. As described in chapter 6, masers are seen in the vicinities of recently formed high mass stars. Thus, one would expect to find masers in starburst regions. Many of the OH masers in starburst galaxies possess intrinsic luminosities more than a million times that of the most powerful OH masers in the Milky Way! These masers are referred to as megamasers, because the prefix mega indicates one million. Gigamasers also have been found, and each is at least one thousand times more powerful than the weakest megamaser. We return to the subject of masers near the centers of galaxies in the next chapter.

9.7 Self-Regulated Structure Formation During the Epoch of the Birth of Galaxies

One of the outstanding mysteries is the presence of heavy elements in the oldest known stars. By *heavy element* we mean any element with a mass per nucleus greater than that of helium. Thus, carbon, nitrogen, and oxygen, as well as iron, are all examples of heavy elements. The presence of these heavy elements in the oldest stars constitutes a mystery because hydrogen and helium are the only elements that were produced in abundance in the Big Bang. Thus, we would expect the first stars to have been born with little other than hydrogen and helium in them. However, we can observe main-sequence stars that we know are nearly as old as the Universe itself! It is surprising that none of these stars are composed of only hydrogen and helium, even though they were formed very early in the Universe's history and they have not evolved to the stage that nuclear processes in them make heavy elements.

The existence of heavy elements in all observed stars implies that a very short-lived first generation of stars was born as the first galaxies were forming or even prior to their formation. Given the abundances of heavy elements in the oldest main-sequence stars, the stellar birth and supernova rates during the epoch of galaxy formation must have been high by present standards. Galaxy formation was a starburst phenomenon! The heavy elements formed in the first bursts were then incorporated in the next generation of stars, including the oldest main-sequence stars observed at present.

The oldest known stars are in objects called globular clusters. One is pictured in Figure 9-8. Each globular cluster contains tens of thousands to a million stars and is roughly a 100 light years across. Each visible star in a globular cluster is somewhat less massive than the Sun, a fact that we know from measurements of their spectra and the theoretical results showing that less massive stars emit a higher proportion of their energy at red rather than blue wavelengths than more massive stars. As related in chapter 3, theoretical work shows that low-mass stars spend longer on the main sequence than high-mass stars. Thus, we can infer the age of a globular cluster from the mass of the most massive main-sequence star in it. More massive stars have left the main sequence to evolve into black holes, neutron stars, or white dwarves.

The issue of how globular clusters formed is at the heart of the problem of the birth of structure in the Universe. One possibility is that the globular clusters appeared before the galaxies did and that galaxy formation was a process of hierarchical agglomeration in which the globular clusters were among, or possibly even *were*, the basic building blocks. In an alternative picture, the superwinds of protogalaxies (i.e., objects evolving to become galaxies) containing bursts of the first generation of stars induced the creation of globular clusters.

We assume that the superwind of one of the young galaxies was similar to but more powerful than that of a starburst galaxy now. Of course, such a superwind blew a bubble having the same basic structure of a windblown bubble as that illustrated in figure 4-6. Intergalactic gas was swept up into a shell after passing through the bubble's leading shock. Theoretical models

FIGURE 9-8   An Optical Image of a Globular Cluster (M80). It is one of 147 globular clusters known in the Milky Way and is at a distance of 28,000 light years. Image courtesy of the Hubble Heritage Team (AURA/STScI/NASA).

show that conditions in the shocked gas were ripe for the formation of molecular hydrogen through two separate routes, both of which operate due to the existence of ionized hydrogen, electrons, and neutral hydrogen together in the shocked swept-up gas. These routes are

$$H^+ + H \rightarrow H_2^+ + photon$$
$$H_2^+ + H \rightarrow H_2 + H^+$$

and

$$H + e \rightarrow H^- + photon$$
$$H^- + H \rightarrow H_2 + e$$

Here e symbolizes an electron. The emission of a photon in the first reaction in each route is a consequence of the requirement that both energy and momentum are conserved in a reaction.

The H_2 was important because it served as a coolant. Just as transitions between bound states in an atom are discrete, the rotational levels of a molecule are separated by discrete energies. The energy separations between rotational levels are usually much less than those between different electronic states of an atom. Thus, collisions involving much less energy than usually required to cause electronic excitation can cause a molecule to become rotationally excited. If a rotationally excited molecule decreases its rotation rate by the emission of a photon, energy is removed from the gas. Collisionally induced excitation of H_2 rotation and subsequent radiation cools gas containing H_2 to much lower temperatures (about 100 K) than those (about 10,000 K) to which purely atomic hydrogen gas would cool in many astronomical environments.

Cooling, resulting from the presence of H_2 in the swept-up gas, lowered the minimum mass for an object to collapse due to its own gravity. Theoretical results imply that that minimum mass was in the range of the masses of the present day globular clusters. These objects may well be a consequence of powerful wind-intergalactic medium interactions in the Early Universe.

SELECTED REFERENCES

Blitz, L., Spergel, D. N., Teuben, P. J., Hartmann, D., and Burton, W. B. "High-Velocity Clouds: Building Blocks of the Local Group." *Astrophysical Journal* 514 (1999): 818–843.

Booth, R. S., and Aalto, S. "Molecular Gas, Starbursts and Active Galactic Nuclei." In *The Molecular Astrophysics of Star and Galaxies*, ed. T. W. Hartquist and D. A. Williams, 437–467. Oxford: Clarendon Press, 1998.

Bregman, J. N. "The Galactic Fountain of High-Velocity Clouds." *Astrophysical Journal* 236 (1980): 577–591.

Bregman, J. N., and Houck, J. C. "The Hot Gas Surrounding the Edge-on Galaxy NGC 891." *Astrophysical Journal* 485 (1997): 159–166.

Cen, R., Phelps, S., Miranda-Escude, J., and Ostriker, J. P. "On the Clustering of Lyman Alpha Clouds, High-Redshift Galaxies, and Underlying Mass." *Astrophysical Journal* 496 (1998): 577–585.

Chevalier, R. A., and Clegg, A. W. "Wind from a Starburst Galaxy Nucleus." *Nature* 317 (1985): 44–45.

Hartquist, T. W., and Williams, D. A. *The Chemically Controlled Cosmos.* Cambridge: Cambridge University Press, 1995.

Hartquist, T. W., Dyson, J. E., and Williams, R. J. R. "Mass Injection Rates Due to Supernovae and Cloud Evaporation in Starburst Superwinds." *Astrophysical Journal* 482 (1997): 182–185.

Jokipii, J. R., and Morfill, G. E. "Ultra-High Energy Cosmic Rays in a Galactic Wind and Its Termination Shock." *Astrophysical Journal* 312 (1987): 170–177.

Kang, H., Shapiro, P. R., Fall, S. M., and Rees, M. J. "Radiative Shocks Inside Protogalaxies and the Origin of Globular Clusters." *Astrophysical Journal* 363 (1990): 488–498.

Lepp, S., and Stancil, P. C. "Molecules in the Early Universe and Primordial Structure Formation." In *The Molecular Astrophysics of Stars and Galaxies*, ed. T. W. Hartquist and D. A. Williams, 37–52. Oxford: Clarendon Press, 1998.

Nagano, M., and Watson, A. A. "Observations and Implications of the Ultra High Energy Cosmic Rays." *Reviews of Modern Physics* 72 (2000): 689–732.

Oey, M. S., and Clarke, C. J. "The Superbubble Size Distribution in the Interstellar Medium of Galaxies." *Monthly Notices of the Royal Astronomical Society* 289 (1997): 570–588.

Savage, B. D., Sembach, K. R., Jenkins, E. B., Shull, J. M., York, D. G., Sonneborn, G., Moos, H. W., Friedman, S. D., Green, J. C., Oegerle, W. R., Blair, W. P., Kruk, J. W., and Murphy, E. M. "Far Ultraviolet Spectroscopic Explorer Observations of OVI in the Galactic Halo." *Astrophysical Journal* 538 (2000): L27–L30.

Shapiro, P. R., and Benjamin, R. A. "New Results Concerning the Galactic Fountain." *Publications of the Astronomical Society of the Pacific* 103 (1991): 923–927.

Snowdon, S. L., Egger, R., Finkbeiner, D. P., Freyberg, M. J., and Plucinsky, P. P. "Progress on Establishing the Spatial Distribution of Material Responsible for the ¼ keV Soft X-ray Diffuse Background Local and Halo Components." *Astrophysical Journal* 493 (1987): 715–729.

Snowden, S. L., Mebold, U., Hirth, W., Herbstmeier, U., and Schmitt, J.H.H.M. "ROSAT Detection of an X-ray Shadow in the ¼ keV Diffuse Background in the Draco Nebula." *Science* 252 (1991): 1529–1532.

Viegas, S. M., Gruenwald, R., and de Carvalho, R. R., eds. *Young Galaxies and QSO Absorption Line Systems.* San Francisco: Astronomical Society of the Pacific, 1997.

10

ACTIVE GALAXIES AND THEIR NUCLEI

At a conference held in Manchester in 1984, Bernard Pagel aptly paraphrased George Orwell by commenting that "all galaxies are active, but some are more active than others." Activity in galaxies manifests itself in a variety of ways. For example, in section 9.6, we described the starburst phenomenon that can lead to the generation of superwinds. In this chapter, we will concentrate on activity (including winds and jets) occurring in the nuclear regions lying within several hundred light years or so of the centers of the galaxies.

The activity in which we are interested is that which cannot be attributed to stars alone. However, we will see that stars can, and indeed, must, be involved. The most obvious manifestation of the activity is the very strong emission of radiation at wavelengths in the X-ray, ultraviolet, and infrared regions and sometimes in the radio region. Very energetic photons in the gamma-ray region are sometimes observable, and as found recently, at least a few rather special objects emit gamma-ray photons that are so energetic that they each carry a million million times as much energy as does a photon in the visible part of the spectrum. The spectrum of a truly active region in a galaxy is very distinct from that emitted by any known star or any ensemble of stars. The spectrum of the kind to which we refer is sketched in figure 10-1. The power of the radiation from an active region ranges from about a few percent of that radiated by all the stars in our Galaxy to 100 times or so greater than the Galaxy's power. The spectra of active regions roughly divide the sources into two groups: objects that are strong emitters at radio wavelengths or *radio-loud* and those that are called *radio-quiet*, though even many radio-quiet objects have detectable radio emission. Superposed on the continuous spectrum is a plethora of emission and absorption lines in roughly the visible through to the near X-ray region of the spectrum (cf. figure 10-1).

The active nuclear region of a galaxy is usually known as an AGN (*active galactic nucleus*). The classification of AGNs can be confusing because it can be far from clear whether objects that have rather different observed properties are in reality intrinsically different. Alternatively, they could be of the same type but in different stages of their evolution, or at the same evolutionary stage, but viewed differently. A present major thrust in AGN studies is the development of "unified models" of AGNs in which differences in properties are related to the orientations of the galaxies with respect to an observer on Earth.

There is a whole "zoo" of AGNs and the denizens include Seyfert galaxies, QSOs (*quasistellar objects*, which include the famous quasars), radio gal-

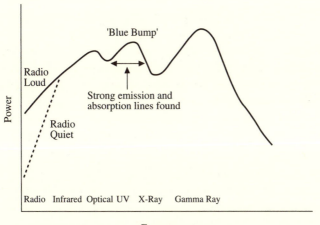

Frequency

FIGURE 10-1   The Spectrum of an AGN from Very Low to Very High Frequencies. The distribution of the average radiated power (per unit interval of frequency) as a function of frequency for AGNs. (The logarithms of the power and frequency are actually plotted.) The solid line represents radio-loud objects, the dotted line represents radio-quiet objects. Note the strong "blue bump" (cf. the very end of section 10.3) in the ultraviolet region. The wavelength region in which strong emission lines and absorption lines are found is indicated. The division into the various spectral regions is appropriate for an observer moving with the AGN (i.e., in the AGN rest-frame). An observer on Earth would see the whole spectrum systematically red shifted.

axies, BL Lac objects, OVVs (*optically violently variable* objects) and LINERs (*low-ionization nuclear emission-line region* galaxies).

We will concentrate mostly on the two largest subclasses, Seyfert galaxies and QSOs, though the distinction between them is, to some extent, one of semantics. However, there are some differences between them. For example, the Seyfert phenomenon is clearly less energetic than the QSO phenomenon. Further, the Seyfert phenomenon is directly observed to be associated with galaxies; the direct evidence for the association of QSOs with galaxies is weaker.

10.1 Seyfert Galaxies

Although the first optical spectrum of an active galaxy (NGC 1068) was obtained in the very early years of the last century, the existence of a whole class of active galaxies was first recognised by Carl Seyfert in 1943. At first glance, a Seyfert galaxy is a normal spiral galaxy with its main distinguishing feature being a highly concentrated bright central core. Plate 9 shows a Seyfert galaxy. A Seyfert galaxy's spectrum shows that the gas has two distinct components. One component produces (relatively) narrow spectral lines emitted by gas having speeds of up to a few hundred kilometers per second and is known as the NLR (Narrow Line Region). The other component has much broader spectral lines emitted by gas having speeds of thousands of kilometers per second and is known as the BLR (Broad Line Region). The emission lines can also be used to deduce the physical conditions in the emitting gas, and they indicate clearly that the BLR gas is heated primarily by a nonstellar radiation field like that sketched in figure 10-1. The NLR is also heated by the radiation field, but there is evidence to show that at least some of the NLR gas emits because it is heated by shock waves. The temperature of the gas in both regions is about the same, several tens of thousands of degrees Kelvin. The most striking difference is that the BLR gas is much denser than the NLR gas. The former has a number density of roughly 10 billion particles per cubic centimeter; the latter has a number density of around 1,000 particles per cubic centimeter.

An important conclusion can be reached from the data. The nuclear volume is at most one thousandth or so of the volume of the entire galaxy and yet is producing, by some nonstellar means, radiative energy at a rate that in some cases exceeds that of all the stars in the galaxy. In fact, some arguments suggest that whatever produces the radiation field is even more centrally condensed than mentioned above. These arguments are based on observations that show that the radiation field varies over times as short as hours. Figure 10-2 shows how the variation time is related to the "size" of the emitting region. So variations of hours would imply a source "size" of around one tenth of a thousandth of a light year! The very high gas veloci-

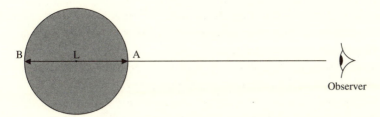

FIGURE 10-2   How Time Variations in Signals Can Lead to Source Size Estimates. The size of a source of radiation can be estimated from the characteristic variability time, Δt, seen in the source signal. If A and B are separated by a distance L, the source size, then changes in the emission from A and B will be seen by a distant observer to be separated in time by $\Delta t \approx L/c$, where c is the speed of light.

ties provide additional evidence that the nuclear region is small. The simplest way to get high velocities is to have the gas in orbit around a gravitating body. Even if the gas is only one light year away from the body, it must have a mass of a billion solar masses to account for an orbital speed of a few thousand kilometer per second.

In 1974 Ye. Khachikian and D. Weedman discovered that Seyfert galaxies can be divided into two classes. A Seyfert 1 galaxy has a spectrum indicating that both a BLR *and* an NLR are present. In contrast, a Seyfert 2 galaxy appears to possess only an NLR. We will see later that this simple classification scheme has been a great stimulus to the developers of unified models.

In the case of an AGN that is not too far away, observations show that the BLR and NLR are spatially very distinct. The BLR is confined to distances of between one thousandth of a light year (Seyfert) and one light year (QSO) from the very central regions of the galaxy. The NLR extends out to distances of hundreds of light years. We would expect this spatial separation from an examination of the physical state of gas exposed to the type of radiation field that is present. The relative abundances of different ions (e.g., O^+, O^{2+}) in the BLR and NLR are rather similar. Yet the BLR is much denser than the NLR. For the relative populations of ions to be similar, the BLR must be much closer to the central radiation source than the NLR. The factor by which the BLR must be closer to the radiation source is roughly equal to the square root of the ratio of gas densities in the BLR and NLR. This factor is about one tenth of a thousandth, consistent, more or less, with the extents of the two regions given above.

10.2 Quasistellar Objects

Quasistellar objects, like Seyfert galaxies, were actually observed, though not appreciated for what they are, rather earlier than generally thought. In the

1930s the Swiss fluid dynamicist-turned-astronomer Fritz Zwicky identified a group of blue, very compact galaxies, some of which are so compact as to appear star-like. Many of the more resolved members of this group have subsequently been identified as Seyfert galaxies; some of the less resolved (or unresolved) members have turned out to be QSOs.

The interest in these objects grew tremendously in the 1960s when the relatively new science of radio astronomy was beginning to open up a golden era in astronomy. Radio astronomers found a plethora of intense radio sources that could sometimes be identified with optically visible galaxies but sometimes, rather puzzlingly, with objects that are of stellar appearance. These latter came to be known as quasars (quasistellar radio sources), one of the most famous acronyms ever invented in astronomy.

In 1963 astronomers combined classical astronomy and modern radio astronomy when they used lunar occultation to identify unambiguously the very strong radio source 3C 273 (a quasar) with a star-like optically visible object. (In this technique, the precise time that the moon starts to block radiation from a source is measured and this helps locate the source.) Maarten Schmidt made the stunning realisation that the only way of understanding the optical spectrum of 3C 273 was for the entire spectrum to be redshifted (cf. section 1.3) by a very large factor. Indeed the factor is so large that the redshift can only be due to the overall expansion of the Universe that is expressed by the Hubble relationship between the distance and speed of an object (cf. section 8.2). 3C 273 turned out to be much further away than any Seyfert galaxy. Perhaps even more startling, when the distance of this object is taken into account, one recognizes that the central source is producing energy at a rate about a hundred times or so greater than a typical spiral galaxy, such as our own Milky Way. That high rate is now known to be typical of QSOs.

Since 1963, many more similar objects have been found; some are so distant that their light has taken almost the age of the Universe to reach us. It has also become clear that about 90 percent of QSOs do not have strong radio emission and so quasars are a relatively small subset of QSOs. The line spectra of QSOs are similar to those of Seyfert galaxies, though there are some important differences. QSO spectra often show very narrow absorption features in heavy elements, such as oxygen and nitrogen, as well as in the Lyman-alpha line of hydrogen (cf. section 9.3). The narrow absorption lines are generally regarded as arising in material lying between our galaxy and these very distant objects. The heavy element absorption lines are formed in the outer halos of distant galaxies whereas the hydrogen absorption lines are formed in clouds of intergalactic gas.

Much more puzzling are the very broad absorption line features seen in the spectra of about 10 percent or so of radio-quiet QSOs. They are *never* seen in quasar spectra. The broad lines cannot be accounted for by intervening galaxies and must be produced in gas physically associated with the QSOs themselves. The gas speeds are sometimes a substantial fraction of the speed of light. Moreover, the lines are always blue-shifted relative to the BLR emission lines. The absorbing gas must be moving toward us (in the QSO frame of reference) at very high speeds. At present no satisfactory explanation for the production of these absorption lines exists, though researchers have conjectured that very fast turbulent winds in the QSO nuclear regions cause the absorption.

10.3 Models for AGNs

The restrictions on any AGN model are severe. Enormous energy has to be produced in a very small fraction of the volume of a galaxy, and this small volume must also contain a mass enormously greater than that of the Sun. We will here describe what is probably best defined as the *default model* for AGNs and discuss later how some of the observational characteristics of AGNs can be explained in terms of this model.

Most astronomers at present regard the Black Hole (BH) model for AGNs as being the most plausible. It explains many aspects of the Seyfert-QSO phenomena, even if watertight proof that this is the correct model does not presently exist.

In 1964, the Russian astrophysicists Ya. B. Zel'dovich and I. D. Novikov proposed, simultaneously with but independently of the American astrophysicist E. E. Salpeter, that AGNs must be powered by BHs. Several strong arguments favor this proposition. First, as mentioned in section 10.1, a single massive object could account for the very high speeds observed simply as a result of its gravity. Second, if material falls onto a BH, the efficiency of conversion of the potential energy of the infalling material into radiation can be relatively high; for example it can be appreciably higher than the efficiency of the nuclear burning processes that produce the energy of main-sequence stars. Third, the sizes of the radiation sources deduced from the time variability arguments are at least consistent with the sizes of BHs with masses hundreds of millions times that of the Sun. Here the size of a BH is the distance of closest approach that light can pass to the BH and escape; for a nonrotating black hole, the so-called Schwarzschild radius is defined to be just infinitesimally smaller than that distance. Finally, examination of the distribution of the brightness in the very central regions of galaxies with active nuclei shows that some of them have very high peaked central brightnesses. This would be consistent with stars gravitationally clustering around a central condensed massive object.

Paradoxically one problem with even such a skeleton model encourages cautious optimism that the model may have some basis in truth. Material a long way away from a gravitating body will not in general fall straight down onto the surface of BH. This is because the material will possess angular momentum due to its motion around that gravitating body. Suppose, for example, that the material falling onto a BH is shed from stars orbiting it. The material cannot fall directly onto the BH but will form an accretion disk around it in which torques produced by friction can remove angular momentum and allow gas to fall from the inside of the accretion disk onto the BH. (Accretion disks are well known to exist around recently formed stars for exactly the same reason; cf. section 5.6.) These accretion disks are hot because they are heated by the friction. The disks have surface temperatures rather similar to the surface temperatures of massive stars, that is, several tens of thousands of degrees Kelvin. They therefore radiate ultraviolet photons producing the very enhanced UV radiation (usually referred to as the *blue bump*) seen in the continuous spectra of Seyferts and QSOs (cf. figure 10-1). The sizes of these disks are inferred to be comparable to the extent of the BLR.

10.4 The Broad Line Regions of Seyfert Galaxies and QSOs

At the time of writing, there is no real consensus as to what the "correct" explanation for the origin of the BLR is. Any acceptable explanation must account for the BLR's properties, which are summarized as follows:

The gas is moving very fast (at several thousand kilometers per second). Perhaps suggestively, its density is rather similar to that found in the outer envelopes of red giant stars and that expected in the surface regions of accretion disks. Given the density, one can infer the mass of gas in the BLR by measuring the amount of light detected in any one of the many spectral features of hydrogen. In a QSO, the BLR mass is up to 100 times the mass of the Sun, but in a Seyfert galaxy it is considerably less. We can work out from the BLR mass, the BLR density, and the volume over which BLR material is distributed what fraction of that volume is occupied by BLR material. The BLR gas fills only one millionth or so of the nuclear volume.

At this juncture, we divide possible models for the BLR into two rather broad classes. First, we have a class in which the BLR is produced by a single object that occupies one millionth of the nuclear volume. The second class is one in which many small units of emitting gas are distributed throughout the nuclear regions but occupy in total one millionth of the available volume. This second class is usually referred to as a "cloud" model.

10.4.1 The Disk Model

As far as the first class of models is concerned, an obvious candidate for BLR production is the accretion disk itself. We have noted that the surface temperature is several tens of thousands of degrees Kelvin; the density in the disk is expected to be similar to that of the BLR gas. Accretion disks rotate and the rotational speed of a disk is roughly equal to the orbital speed of gas around the central massive body. Thus, the surface regions of a rotating accretion disk could be the sites of the BLR. A slightly different version of this model follows from the comment made in section 10.3 that accretion disks have surface temperatures similar to those of massive stars. In section 6.3 we described radiatively driven winds from massive stars, and one might reasonably expect that winds could be driven from the accretion disk surfaces, producing high-speed gas. Provided most of the emission comes from that portion of the wind closest to the disk, the differences between the two models are small. We will return to the accretion disks in section 10.4.4.

10.4.2 The Red Giant Atmosphere Model

In the second class of models, an obvious candidate for a cloud is the atmosphere of a red giant star that is heated by the central radiation field, since, as we have noted, the BLR densities are similar to those found in such stellar atmospheres. Several million red giant stars would need to be present to account for the BLR. The orbital speeds of the stars around the central BH would produce the BLR line widths. It is worthwhile going into this particular model a little more to show how sometimes what appears at first sight to be a plausible model can be undermined, once it is investigated in more detail.

Very roughly, if we randomly selected 10,000 stars in a galaxy, only one of them would be a red giant. Thus, if we need a million red giants to account for the BLR, we would expect in total about 10 billion stars to be present. They would have an enormous mass (even greater than that estimated for a central massive object) and an enormous space density, making a red giant model for BLR improbable. Further, since the stars must move around the region occupied by the BLR, collisions between stars would occur and any giant star would experience a collision that would destroy its atmosphere in an embarrassingly short time. This collision time is so short that even if an isolated giant star could reestablish its extended atmosphere after it was removed, the next collision would occur before the atmosphere could be reestablished. The red giant model seems now rather implausible.

10.4.3 The Pressure-Confined Cloud Model

If the red giant model is deficient, then we have to look for some other type of clouds. To go further, we have to look at the concept of isolated clouds in more detail and examine the question of how long an isolated cloud might exist. Unless they are held together by some force, clouds simply expand. What forces could keep them together? Stars, of course, are held together by their self-gravity (another attraction of the model of section 10.4.2!). However, the self-gravity of BLR clouds is simply far too small to do this. There are two other possibilities. One is to invoke a magnetic field around a cloud to stop it from expanding. Unfortunately, although magnetic fields can prevent gas expanding across the field lines, they are less effective in stopping gas from expanding along the field lines. The second possibility is to have some gas outside the cloud that exerts a pressure on its surface. We will explore this possibility in some detail, since it provides an excellent example of how astrophysical theories have to be modified in the light of observational discoveries and theoretical arguments.

Suppose a spherical cloud of a certain pressure is prevented from expanding by a surrounding medium; the surrounding medium clearly must have the same pressure as the cloud. We have seen that the pressure of a gas is proportional to the product of its number density and temperature. A confining medium that has the same density and temperature as the cloud is not reasonable because it would then be indistinguishable from the cloud. Either the confining medium is cooler and denser than the cloud or it is hotter and less dense than the cloud. Surrounding cool gas would obscure the cloud; therefore, the confining medium must be hotter than the cloud.

The physical picture we are presenting is rather like that of a plum pudding. The cooler BLR clouds are the "plums" embedded in hotter confining gas that is the pudding itself. We need to look for the production mechanism for this system, that is, the recipe. To do so, we now go back to our comment in section 10.1 that the central radiation field heats the BLR gas. 3C 273 is a particularly noteworthy object, not only because it was the first quasar to be unambiguously identified with its optically visible counterpart. It also was one of the first QSOs whose entire spectral energy distribution was determined with any accuracy.

The spectral energy distribution was used in calculations of the temperatures of gas heated by AGN radiation fields. In a landmark paper published in 1981, J. H. Krolik, C. F. McKee, and C. B. Tarter showed that at the pressure deduced for the cool BLR clouds, the gas could exist in two very distinct phases. One was at the inferred BLR gas temperature and density. The second was at a much higher temperature of about 100 million degrees Kelvin and at a density about one tenth of a thousandth that of the cooler gas. Thus, the coexistence of BLR gas and a hot confining medium at the same pressure, as required in the plum-pudding model, seemed likely.

Closer examination unfortunately reveals serious problems with this model. First, we know the BLR clouds have to move at several thousands of kilometers per second and they would therefore be moving relative to the hot surrounding gas. The relative motion produces small pressure differences on the surface of the clouds, causing them to evaporate from the lowest pressure parts of their surfaces (cf. section 8.5).

A second serious problem arose when it was discovered that the spectrum of 3C 273 is atypical of QSO and Seyfert spectra; it has a much greater proportion of photons at higher photon energies than usual. For a more typical spectrum, calculations show that the hot phase has a temperature of about 10 million degrees Kelvin instead of the 100 million degrees Kelvin found using the spectrum of 3C 273. Consequently, to have the same pressure as the BLR clouds, the hot phase must have 10 times the density it would have if the 3C 273 spectrum were appropriate. At first glance this does not seem too serious. The problem is that X-ray observations of Seyfert nuclei show time variability over timescales so short that the X-rays must be generated close to the central BH. At the higher hot phase densities implied by the lower temperature, there would be so many electrons around that they would act like a fog and would obscure the X-ray emitting regions so much that the time variability would not be seen.

In fact, calculations based on newer measurements of AGN spectra go even further in destroying the two-phase plum-pudding model. Although somewhat dependent on the assumptions made (e.g., for the relative abundances of elements), most calculated results show that a clear-cut, two-phase structure does not even exist for a pressure typical of that of the BLR.

10.4.4 An Evanescent Cloud Model

Since it seems impossible to confine clouds we have to abandon the idea that clouds have long lives. We suppose instead that they are produced at a rate that is about equal to the rate at which they are destroyed by expansion. Such a cloud population is called *evanescent*. An individual cloud has a short existence, but the cloud system itself is maintained due to the production.

We have already discounted the possibility that red giant atmospheres provide the BLR, but we have not ruled out the presence of stars in the vicinity of a BLR. Since Seyfert galaxies are galaxies, it would be surprising if stars were *not* present in the Seyfert nuclear regions. The evidence that QSOs are associated with galaxies is not so clear-cut, but there is enough evidence (at least for the less distant QSOs) for the proposition that they are to seem plausible.

A much stronger argument can be advanced for the presence of stars in AGNs. The BLR gas contains heavy elements and only nuclear burning in stars can produce them. The BLR must be composed of mass lost from stars. Stellar mass loss may fuel the central AGN engine to produce the radiation field.

Little is known about star formation in the centers of galaxies. However, if star formation took place in the center of a galaxy, stars with a whole range of masses may form just as stars with a range of masses form throughout the Milky Way. In particular, the massive short-lived stars that evolve into supernovae might be born in a galactic center. The blast waves of the supernovae would overrun surrounding material and both the stellar ejecta and surrounding material would be shock heated to very high temperatures. The temperatures produced in the early stages of interaction can be far higher than those of a hot phase heated by the central radiation field. The fact that the temperatures produced in the shocks are so high is, curiously, the key to how much cooler clouds can be produced.

The laws of thermodynamics tell us that in an isolated system, heat flows from a hotter to a colder body. In this case, the shock-heated gas is the hotter body and the radiation field is the colder "body." The radiation field picks up energy from the gas and the gas cools. Basically, the photons are just scattered by electrons in a process called *Compton scattering*. The process takes place gradually enough that the gas pressure remains almost constant. Therefore, as the gas cools, its density must increase to maintain the pressure. If Compton scattering were all that happened, the gas would end up at the hot-phase temperature of 10 million K and cool BLR clouds would not be produced. However, as the gas cools, other cooling processes come into play (cf. the comments in section 8.4 on cooling mechanisms). The gas temperature then plummets below the hot phase temperature and cool gas is produced. During the cooling, clouds are produced because bits of gas that have cooled initially just a little more than their neighboring bits continue to cool much faster than the neighbors do. Thus, provided supernovae keep going off in the nuclear regions, BLR clouds can be continuously produced. In the supernova induced cloud formation picture, some of the BLR speeds are produced because the cooled material is either fast moving gas ejected by supernovae or gas that has been set into motion by being overrun by the supernovae induced blast waves. The pictures seem appropriate for QSOs. One suggestive piece of evidence in its favor is that the supernova rates needed to produce the estimated mass of BLR gas generate enough material to account for the observed QSO luminosities, if roughly half the material produces the BLR and the other half powers the QSO by falling into the BH.

The births of BLR clouds in Seyfert galaxies are probably not induced by supernovae. The nuclear regions of Seyfert galaxies are much smaller than those of QSOs, and a single supernova would blow gas away from the entire nuclear region. However, massive stars may, nonetheless, play a role in BLR cloud formation in Seyferts as their stellar winds are powerful but not so powerful that they will necessarily blow all the gas from a Seyfert nucleus. Cooling of gas behind stellar wind-driven shocks may be the mechanism for BLR cloud formation in Seyfert galaxies.

We noted at the end of section 10.4.1 that we would return to the disk model. The BLR spectra show that a BLR most probably has two components. The data might be explained by the combination of the accretion disk model with the evanescent cloud model. The accretion disk can produce the various hydrogen lines; the clouds can produce the emission from more highly ionized species (such as trebly ionized carbon). Figure 10-3 is a sketch of the main components of the BLR region in the combination model, though we shall see later that some modifications to this picture are necessary.

10.5 The Narrow Line Regions of Seyfert Galaxies and QSOs

Like the BLRs of the Seyferts and the QSOs, the NLRs of these two types of AGNs are remarkably similar. We will concentrate here on Seyfert galaxies for the simple reason that their NLRs can be resolved; in other words, they are big enough and near enough to be seen clearly. Thus, far more is known about NLRs in these sources. An important reason for studying an NLR is that if the fuelling of the central engine takes place as a result of gas infalling from the galaxy as a whole (as opposed to stellar mass loss in the very inner nuclear regions where the BLR is formed), then we can gain clues about the fuelling process from an analysis of the NLR structure.

We have assumed a gravitational origin for the BLR speeds to argue for the existence of a massive central body. We need not speculate that gravitational forces play some role in the NLR. Direct observational evidence confirms that they do. Because the region is resolved, it is relatively straightforward to measure the typical speeds of stars in an NLR. The stellar speeds have to be of gravitational origin. Since there is a strong direct correlation between them and the NLR spectral line widths, the gas motions also have to be governed partly by gravity. However, some evidence suggests that additional influences are at work. Data indicate that as the radio power from the nuclear regions

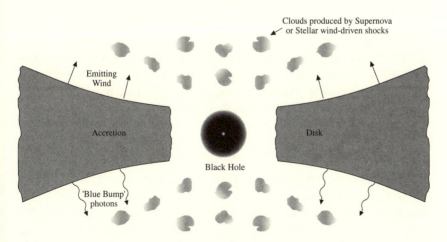

FIGURE 10-3   A Sketch of the Main Components in the Very Inner Regions of an AGN. The sketch is not to scale. This inner region extends out to the BLR scale (see text).

increases, so do the NLR line widths and line strengths. We shall return to this point later.

The same types of argument used to estimate the fraction of a BLR filled by gas imply that the fraction of a NLR filled by gas can be as low as that of the BLR but generally is appreciably higher. Typically about 1 percent or so of the region is occupied by gas.

We cannot invoke an accretion disk to explain the origin of the NLR, since an accretion disk is small and is located in the same region as the BLR. The density of the NLR gas is orders of magnitude below that of the BLR, and a stellar origin is improbable.

A promising alternative explanation is based on the comment made above concerning indications that the NLR gas properties have some connection with the radio power. Maps show close spatial associations of features in the NLR detected in the optical and structures apparent at radio wavelengths. Radio emission arises in and around jets or localised regions of hot plasma (sometimes called plasmons) which might originate in the nuclear regions near the BH. Jets sweep through local gas and accelerate it via shock waves; similarly, plasmons can expand and sweep up gas by shocks just as supernova remnants do (cf. section 8.4).

Any material swept up and shocked by an expanding plasmon has to cool if it is to form NLR clouds. Cooling takes place exactly in the same way as it does behind the shock in an expanding supernova remnant. Radiative cooling behind supernova remnant shocks becomes important when the supernova shock speed drops to about a few hundred kilometers per second, which is also the typical shock speed for the onset of radiative cooling in gas shocked by an expanding plasmon. Gas must cool radiatively to be prominent in optical emission. Thus, a single expanding plasmon will contain optically detectable material with a range of line-of-sight velocities of at most 500 or 600 km s^{-1}. This is remarkably similar to the velocities seen in the NLR.

The ambient material that is swept up may be gas that has been evaporated off a large-scale molecular structure by, for example, heating by the central radiation field. The role that such structures play in unified models of AGN are examined in section 10.7.

10.6 Cool Gas in AGNs

So far, we have concentrated on the gas that comprises the BLR and NLR of an AGN. The gas detected in optical emission has temperatures of tens of thousands of degrees Kelvin. Much cooler gas in the form of neutral atoms and molecules (as opposed to fully ionized plasma) exists in AGNs. This gas has a temperature of perhaps 100 K or even less. CO, H_2O, and OH produce emission lines in the cool gas. The latter two molecules yield particularly bright emission because they radiate maser features (cf. section 6.2); the emission can be so bright that its sources are often megamasers (cf. section 9.6). The molecular gas must contain dust, since its role as a catalysis is necessary for molecule formation.

CO observations of nearby galaxies containing AGNs show that the molecular gas in each is distributed in a torus (or ring, or disk), which has a size of several hundred light years and is situated around the central energy source. The gas is rotating at about 200 km s^{-1} around this source. If the rotation is caused by a central object, it must have a mass of several billion times that of the Sun, which is consistent with the mass required to explain the BLR speeds if they have a gravitational origin.

Many Seyfert galaxies show absorption in atomic hydrogen and OH. The absorption data imply the existence of the same sort of gaseous structure as is detected in the CO emission.

OH megamasers also indicate the presence of molecular tori with sizes of tens of light years. The OH tori are probably the inner regions of the larger scale structures detected in CO emission. The water megamasers, however, are rather different. They are much closer to the central sources and show the presence of much smaller structures with characteristic sizes of between about one third to one light year. This is comparable with the sizes expected for accretion disks, and the water maser motions are compatible with their being in a disk.

The shape of what is called the Extended Narrow Line Region (ENLR) provides further evidence for large toroidal structures. The ENLR is the source of emission from ionized gas that extends over distances of thousands of light years, far beyond the NLR. Some of the radiation around the central BH escapes through both the BLR and the NLR to heat and ionize the ENLR gas. The distribution of ENLR gas indicates that ionizing radiation from the AGN central regions escapes into conically shaped regions, as would be expected if the central BH is surrounded by a large-scale torus as depicted in figure 10-4.

10.7 Unified Models of AGNs

A large-scale cold molecular torus exists on the scale of the NLR. An accretion disk exists on the scale of the BLR. A BH is at the center. This system is sketched in figure 10-4. The most exciting aspect of the sketch is that it suggests a unified model for AGNs. As is seen in figure 10-4, the division of Seyfert galaxies into Seyfert 1s and Seyfert 2s can be explained as an effect caused by the orientation of the AGN with respect to the observer. An observer sees down into the very central regions of the AGN in a Seyfert 1 and can detect both the NLR and the BLR. An observer's line of sight to the BLR in a Seyfert 2 is blocked by the large torus and only the NLR is seen.

As indicated in figure 10-4, we might detect faint BLR emission even if the direct line of sight into the BLR is blocked, provided that BLR photons

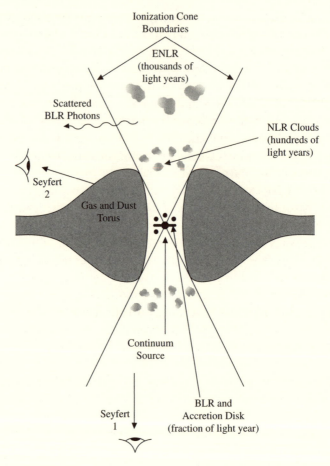

FIGURE 10-4   The Main Structures in an AGN. The sketch is not to scale. The sketch shows the main features that lead to a unified model for Seyfert galaxies. A large-scale torus of cool gas surrounds the BH/accretion disk inner system (figure 10-3). The three emission regions (BLR, NLR, ENLR—see text) are indicated. The ionization cones that allow the ionization of the ENLR are shown. A possible scattering path for BLR photons into a Seyfert 2 line-of-sight is shown.

are scattered off dust or electrons into the observer's line of sight. Such faint emission was first detected in the Seyfert 2 galaxy NGC 1068 in 1985.

10.8 Radio Jets, Ultra-High-Energy Cosmic Rays, and TeV Gamma-Rays

When examined at radio wavelengths, quasars (as noted earlier, those QSOs that are radio-loud) often are seen to possess huge (tens to hundreds of thousands of light years in extent) radio emitting lobes that are situated on either side of the central galaxy. Plate 10 shows the radio jet of an AGN. Often, a jet of radio emitting material seen can be traced back from the lobe to the central object. The energy required to power the emission from the lobe is channelled there along the jet from the central energy source. The axis of rotation of the BH-accretion disk system provides a natural axis of symmetry for the jet. Quite how the jet is produced and collimated is still a major puzzle, but it is generally believed that strong magnetic fields are in some way involved (cf. section 5.6 in which rotation and magnetic fields are discussed in the context of stellar jets). Magnetic fields also play another role: the observed radio emission from the lobes and jets is synchrotron emission produced by the motions of relativistic electrons in the presence of magnetic fields (cf. section 8.6).

Many, though not all, AGN jets are moving at speeds very close to that of light. Several lines of evidence support this statement. For some jets, motions (on the plane of the sky) of features are detectable through the comparison of radio images made several years apart. The apparent speeds of some features exceed the speed of light. Motion faster than the speed of light is in contradiction to a basic tenet of the Theory of Relativity. An analysis based on that august theory shows that subluminal motions can appear superluminal if two conditions are satisfied. These are that the jet must travel at a speed very close to the speed of light *and* that the jet must be pointing almost directly at the observer. The explanation of the apparent superluminal motions is supported by the observation of one-sided jets, which are seen commonly. It seems very unlikely that jets are intrinsically one-sided. However, if material is moving at a speed close to the speed of light, radiation produced by this material is strongly beamed in the direction of travel. (This phenomenon is called *aberration* or *Doppler boosting*.) If two jets come from a source and one jet points at an observer and the other points away, then the observer would see only the radiation from the jet moving toward him or her.

The relativistic speed, that is, a speed close to that of light, of a jet provides evidence for a BH at the center of an AGN with a relativistic jet. As remarked earlier in section 6.1, the escape speed near a BH would be expected to be a substantial fraction of the speed of light.

Jets associated with BHs with masses of a few times that of the Sun are observed in the Galaxy. These BHs are the end products of some Type II supernovae explosions. They are sometimes referred to as *microquasars*.

In figure 8-6 we showed the complete cosmic ray spectrum measured on Earth with the so-called "knee" and "ankle" well identified. In sections 8.6 and 9.5 we discussed how cosmic rays are accelerated; those up to the knee may be accelerated by the first-order Fermi process in SNRs and those between the knee and the ankle by the same process in the interaction of the Milky Way wind with the intergalactic medium. Shocks can arise in jets as a result of pressure variations around the jets (such shocks can be seen in the exhausts of jet engines), and these shocks are likely to be the sites of particle acceleration. Particles with energies higher than that associated with the knee may be accelerated in the jets of some AGNs.

BL Lacertae (BL Lac) objects are a small but intriguing special group of AGNs. Their spectra have two particular defining characteristics: they have no, or at most very weak, emission lines and they are the only identified extragalactic sources of the very high-energy gamma rays mentioned in the introduction to this chapter. They are very strong radio wave emitters and must have associated relativistic jets. The present interpretation is that one jet in each of the BL Lac objects is pointing more or less directly at the Earth, and the Doppler boosted radiation emitted in the jet simply swamps the emission lines. The very high-energy gamma rays are produced by collisions between very energetic electrons accelerated in the jets and photons, in a manner analogous to the scattering process described in section 10.4.4. However, much higher energies are involved in the production of the very high-energy gamma rays.

10.9 Our Galactic Center (GC)

The nearest galactic nucleus to the Sun is the center of our own Galaxy, at a distance of somewhat under 30,000 light years. Since it is so near, we can study it at far greater spatial resolution than we can the nuclei of external galaxies. However, the GC is observable only at wavelengths beyond the infrared because interstellar dust obscures it. Optical (or shorter wavelength) observations are limited to regions in the plane of the galaxy closer to us than about one third the distance to the GC.

The GC seems to house a massive object, though it is appreciably less massive than the BHs at the centers of the most powerful AGNs. In 1974, a bright compact source of radio radiation known as Sgr A* was discovered at the center of the Galaxy. There are two clusters of evolved high-mass stars orbiting Sgr A*. The stellar motions imply that there is a central condensation of matter of about 3 million solar masses within about 15 light days of Sgr A*. Additional evidence suggests that Sgr A* is massive. Its motion relative to the center of mass of the entire system of stars at the GC can be measured accurately, and it is moving far more slowly than the individual stars. Just as a lot of rapidly moving golf balls hitting a cannonball from random directions do not cause the cannonball to move rapidly, many stars orbiting Sgr A* would not cause it to acquire a high speed if its mass is much higher than that of a star. The low speed of Sgr A* compared to that of the stars implies that its mass is at least 1,000 times that of the Sun. As the minimum mass that it could contain is so high, Sgr A* may well consist of almost all of the 3 million solar masses known to be near its location. It is probably a BH.

Molecular line observations have also shown that a ring of molecular material exists at a distance of a few light years from the GC. Thus, if Sgr A* is a BH, our GC has at least two features in common with an AGN.

The mechanism that produces the radio emission is not fully understood, but it is most likely a by-product of the accretion of material on to Sgr A*. The accreting material probably originates in a concentrated cluster of hot stars known as IRS 16. The cluster lies within the molecular ring. The IRS 16 stars are Wolf-Rayet stars which (as we have discussed in section 7.2) have strong stellar winds. A fraction of the mass lost by the IRS 16 stars falls onto Sgr A*. It is possible to estimate the rate at which material should fall onto Sgr A* if it is a several million mass BH. The rate at which radiation should be produced is estimated from the infall rate. Our GC produces less radiation than expected for a supermassive BH near IRS 16 by a factor estimated to lie between 100 and 1,000! Perhaps, the magnetic field around IRS 16 is very weak, causing the synchrotron emission process giving rise to radio emission to be inefficient.

Not all the wind material from IRS 16 accretes onto Sgr A*. The interaction of this material with another star in the GC has been observed. An infrared source referred to as IRS 7 lies about one light year from the GC and, thus, at about the same distance from IRS 16. IRS 7 has been identified as a red supergiant star (cf. section 3.3.2). Radio maps show that IRS 7 has a long tail of material (rather like a cometary tail) that points away from the GC. The tail is produced as the collective wind from IRS 16 filters through the extended clumpy atmosphere of IRS 7 and strips material off the atmospheric clumps. The stripped material is accelerated by the IRS 16 wind along the direction from IRS 16 to IRS 7.

REFERENCES

Antonucci, R. "Unified Models for Active Galactic Nuclei and Quasars." *Annual Review of Astronomy and Astrophysics* 31 (1993): 473–521.

Kormendy, K., and Richstone, D. "Inward Bound—The Search for Supermassive Black Holes in Galactic Nuclei." *Annual Reviews of Astronomy and Astrophysics* 33 (1995): 581–624.

Krolik, J. E., McKee, C. F., and Tarter, C. B. "Two-phase Models of Quasar Emission Line Regions." *Astrophysical Journal* 249 (1981): 422–442.

Morris, M., ed. *IAU Symposium 136—The Center of the Galaxy*. Dordrecht: Kluwer Academic Publishers, 1989.

Morris, M., and Serabyn, E. "The Galactic Center Environment." *Annual Review of Astronomy and Astrophysics* 24 (1986): 171–203.

Peterson, B. M. *Active Galactic Nuclei*. Cambridge: Cambridge University Press, 1997.

Robson, I. *Active Galactic Nuclei*. Chichester: Praxis Publishing, 1996.

Yusef-Zadeh, F., Melia, F., and Wardle, M. "The Galactic Center: An Interacting System of Unusual Sources." *Science* 287 (2000): 85–91.

11

SOME OTHER WINDY AND EXPLOSIVE SOURCES

Though in previous chapters we have addressed a large variety of astronomical objects and phenomena, we have by no means examined all of the types of environments in which winds and explosions occur. The present chapter contains brief descriptions of several classes of windy and explosive sources upon which we have either touched only briefly or which we have not yet mentioned.

In the closing section of chapter 5 on low-mass young stellar objects, we gave some hint that outflows from such sources are not continuous. However, we did not examine the FU Orionis phenomenon, named after the prototypical source in which it occurs. The FU Orionis phase is one of exceptionally high brightness and mass loss for a LMYSO.

Previously, except in our treatment of Type Ia supernovae, we have neglected the binary stellar systems in favor of single stars. However, about half of the known stars are in gravitationally bound pairs. The symbiotic stars are amongst these, and the collision of the winds of the stars in a symbiotic pair is responsible for some of the distinguishing morphological characteristics of the gas surrounding the binary. Some symbiotic stars are associated with "very slow" and "recurrent" novae.

The "classical novae" receive some attention in section 1.3 in which S. R. Pike's important work on them is mentioned. We give further consideration to these explosive sources in this chapter.

For a quarter of a century, the locations and natures of the gamma-ray bursts constituted a major unsolved problem. Some astrophysicists argued that the bursts occurred long ago at distances comparable to the size of the observable Universe, while others maintained that the detected bursts originate in the Milky Way. Observations at wavelengths other than in the gamma-ray part of the spectrum resolved this interesting dispute.

11.1 The FU Orionis Phenomenon

The star FU Orionis is the prototype of a class of sources identified as LMYSOs with large outbursts in their optical luminosities. Increases by factors of several tens to several hundreds over periods of months to years are followed by decreases on timescales of many tens to several hundreds of years. The P-Cygni profiles (cf. section 1.3) of lines of such sources indicate mass loss rates of ten millionths of a solar mass per year.

FU Orionis-like systems have accretion disks (cf. section 5.5). A star has a single characteristic surface temperature. The fact that an accretion disk has a wide range of temperatures on its surface is key to several of the main techniques used to establish observationally the presence of a disk in a source. An accretion disk is heated by friction, which arises because material nearer the central star orbits at a higher speed than material just slightly further away, just as the Earth's orbital speed around the Sun is greater than that of Jupiter. The heating rate per surface area of an accretion disk goes up as the orbital speed and differences in the orbital speeds of material at slightly different locations increase. Thus, the heating is greatest near the axis about which the disk rotates, and the most distant parts of the disk are the coldest. Figure 11-1 shows the temperature of a disk as a function of position in a standard model.

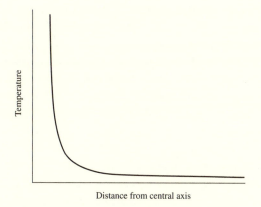

Distance from central axis

FIGURE 11-1   The Relative Temperature of a Disk as a Function of Distance from the Rotation Axis. The curve is for a standard theoretical model.

The spectrum of radiation emitted by an object with a distribution of temperatures differs from that of an object at a single temperature. Infrared and submillimeter observations of FU Orionis-like systems show spectra more compatible with the emission arising from a central star surrounded by a disk with a range of temperatures than from a spherical star with a single characteristic temperature.

Given the existence of observational evidence for the presence of accretion disks in FU Orionis-like systems, the issue of how an instability in a disk might lead to an outburst is relevant. An instability is a runaway phenomenon. A state set up initially is often subjected to a small disturbance. For instance, we might place a ball on a pointed stick in such a way that it initially balances and does not fall due to the Earth's gravity. However, if we disturb the state slightly by blowing on the ball we will ruin the balance. At first the ball will move slowly in the direction that the breath pushed it, but as it moves further from the position in which it was balanced, gravity causes the motion away from that position to get faster and faster until the ball falls completely off the end of the stick.

One mechanism that can lead to a runaway in some accretion disks is called thermal instability. The thermal instability in an accretion disk around a LMYSO may arise as a consequence of the opacity of matter depending on the temperature in a fashion favorable for the onset of instability. Many types of material in many environments cool primarily by the emission of light and other electromagnetic radiation (e.g., infrared and submillimeter radiation). However, such radiation does not always escape easily from an object. The opacity is a measure of the difficulty radiation has in escaping, and it depends on the temperature of the material. The dependence exists because the temperature influences the abundances of various atomic and molecular species and the ways in which their internal structures are excited. The internal excitation affects the efficiency of the interactions of the species with radiation. In some temperature ranges the opacity of disk material increases with temperature. As a consequence, a slight rise in temperature in one part of the disk will make the escape of radiation from it more difficult, which will result in the material becoming hotter, which will increase the opacity and decrease the rate at which radiation escapes, which will make it hotter still.

The runaway heating results in the storage of heat that must be released somehow. Heat may be released due to rapidly developing motions driven by the pressure of the heated region. These motions, resulting as a direct consequence of the thermal instability, may greatly alter the structure and dynamics of an accretion disk dramatically for some time. If thermal instability is the ultimate cause of FU Orionis-like outbursts, it causes the rate of material moving through the disk toward the central star to increase from ten millionths of a solar mass per year by a factor of as much as 100. Such a violent change in disk structure results in a large enhancement of the mass loss rate due to the wind of the star-disk system.

We touched briefly in section 5.7 on evidence for episodic mass loss driving CO outflows and Herbig-Haro object dynamics around LMYSOs. The FU Orionis phase may be one through which nearly all LMYSOs pass several times before evolving into a main-sequence star.

11.2 Symbiotic Stars

Accretion disks with certain characteristics fragment. Such characteristics include a high enough ratio of the energy associated with the rotation to the thermal energy. Fragmentation leads sometimes to the production of binary star systems.

Our treatment of stellar evolution in chapter 3 was focused on single stars rather than members of binaries. Binary membership can affect a star's evolution.

In most binary systems the stars are separated sufficiently that when both are in the main-sequence (cf. section 3.2) phase of evolution, they behave more or less as individual stars. The more massive star of such a system will evolve from the main-sequence phase more rapidly than the less massive one will. The envelope of that more massive star will then extend as much as hundreds of times the star's main-sequence radius (cf. section 3.3). In a binary system in which the stars are close enough together a good fraction of such an extended envelope moves into the region in which the gravity of the less massive, less evolved star dominates. The smaller star gains mass. The initially less massive star sometimes (but not always) becomes more massive than the more evolved star. That such mass transfer occurs in some binary systems led to the resolution of the so-called Algol paradox. The Algol system is one in which the less massive star is the one that is more evolved even though the theory of the evolution of single stars predicts that if two stars of different mass are born simultaneously, the more massive one will evolve more rapidly. It was only in 1955 that J. A. Crawford became the first to offer mass transfer between stars in the binary as an explanation of the relatively greater degree of evolution of the less massive star in Algol.

Many binary systems are referred to as *symbiotic stars*, where *symbiotic* implies the coexistence of two species. Symbiotic stars were identified as a class before it was possible to use telescopes to resolve many of them into two distinct spatially separated images and before they were clearly identified as binary star systems. From the time of the first identification of the class, membership in it was claimed for a system if its spectrum shows the reddish emission of a star in a highly evolved giant phase and emission from highly ionized species not present in an isolated giant star. Now we know that a symbiotic star is really a binary system with an orbital period of years to tens of years and a separation comparable to that between the Sun and one of its planets.

The explanation of the morphologies and dynamics (as inferred from the measurements of redshifts and blueshifts of spectral features) of the ionized gas distributed around symbiotic stars is being developed on the basis of models of the interactions of the winds of the two stars in a binary. Plate 11 shows a photograph of an ionized nebula of a symbiotic system; its crab-like morphology is shared by the distributions of gas around some, but not all, other symbiotic stars. In some wind-wind collision models, one star is assumed to have an outflow similar to that of a low-mass red giant star (cf. section 7.5) whereas the other is assumed to have a wind more like that of the central star of a planetary nebula (cf. section 7.8). One of the stars is taken to be as hot as the central star of a planetary nebula is. The effects of the ultraviolet radiation from the hot star on the winds are taken into account and considered responsible for inducing the production of many observable optical and ultraviolet emission features.

The mechanism driving the wind from the hotter star in a symbiotic system may simply be the action of radiation pressure on matter as in the hot central star of a planetary nebula. However, many symbiotic systems show variations on timescales from tens of years to minutes, and it is possible that the wind arising in the vicinity of the hotter star of a system emanates from a disk surrounding it. As pointed out earlier in this section, mass is transferred between stars in many binary systems; mass transfer leads to the formation of accretion disks. Variability in mass loss from disks was introduced in the previous section. However, variability need not always imply the presence of an accretion disk, as instabilities are known to occur in radiatively driven winds from single stars.

11.3 Novae

Although the stars in a symbiotic pair have typical orbital periods of years, some binaries contain stars that are much closer together and have correspondingly shorter orbital periods. A cataclysmic variable is a binary system with an orbital period of hours and in which one star is a white dwarf and the other is usually a low-mass main-sequence star. The separation between the members is only several times the radius of the Sun.

The stars of a symbiotic pair were initially more distant from one another, but tidal interaction has brought them closer. The most familiar example of tidal interaction is that between the Earth and the Moon causing the ordinary tides in the oceans and seas. Though harder to perceive, tides in the solid component of the Earth and in the Moon also rise and fall. The ultimate source of tidal energy is the energy of the orbital motion and of the rotation of the Earth about its axis. The transfer of the energy from the orbit and the rotation to the tides slow down the Earth's rotation and causes the Moon to fall slowly toward the Earth.

The stars in a cataclysmic variable are sufficiently close to one another that mass is accreted by the white dwarf from the atmosphere of the companion. An accretion disk exists around the white dwarf.

The *classical novae*, mentioned in section 1.3, occur in cataclysmic variables. The brightness of a classical nova rises from the prenova brightness by a factor of up to ten thousand to a million in a couple of days. It typically drops down again by a factor of about two every several weeks. About one tenth of a thousandth of a solar mass of material is ejected in the explosion.

A classical nova results from thermonuclear runaway in material accreted onto the surface of the white dwarf. This thermonuclear runaway differs from that mentioned in the discussion in section 8.1 of Type Ia supernovae in that it occurs in accreted material that is predominantly hydrogen and helium rather than in the carbon-rich material at the *center* of a white dwarf. However, the runaway is similar to that in a Type Ia supernova in that the predominately hydrogen and helium material involved in a classical nova explosion is degenerate, which makes the runaway possible.

At first glance it would seem that the classical novae and Type Ia supernovae occur in somewhat similar types of binary systems. It is possible that some systems in which novae take place eventually give rise to Type Ia supernovae. However, to create a supernova more mass must be accreted and the accretion probably must take place at a much higher rate than in a system in which a nova explodes. The properties of a binary system in which a supernova results must, therefore, differ from those of most cataclysmic variables in which classical novae appear.

Our use of the term *classical nova* rather than *nova* implicitly implies the existence of other types of novae. Other classes include *recurrent novae* and *very slow novae*. In principle there is no difference between a classical nova and a recurrent nova except that the latter has been observed to be in an active nova phase more than once. Some recurrent novae are associated with binary pairs in which one star is a white dwarf and the other is a red giant, with an orbital period of close to a year; such a pair more closely resembles a symbiotic system than a cataclysmic variable. A *very slow nova* is characterized by having undergone only one observed eruption with an increase in brightness of a factor of up to about 300 and a decay time of hundreds of days rather than the tens of days associated with decay of a classical nova. The very slow novae occur in systems that appear to be symbiotic.

11.4 Gamma-Ray Bursts

Gamma-ray bursts were first reported in 1973. They were detected with a satellite designed to observe the sky at wavelengths of around one millionth of a micron, which is about one million times shorter than the wavelength of red light and near infrared radiation. Bursts last from about several hundredths of a second to 100 seconds. Often a single one appears brighter at gamma-ray energies during its short life than all other gamma-ray sources added together.

By 1995, over 2,000 gamma-ray bursts had been recorded, but so little was known of them that at that time debate continued over whether they originate in the Milky Way or are primarily at distances comparable to the extent of the observable Universe. The main problem in identifying the locations of the bursts was that gamma-ray detectors do not have high directional resolution. The independence of bursts' frequency of occurrence with direction in the sky means that they are not confined to the disk of the Milky Way; advocates of a Galactic origin of the sources argued that the gamma-ray bursts occur in the halo of the Galaxy, which they supposed to extend to distances more than ten times the distance of the Sun to the Galactic Center.

Whether the sources are in the halo or at cosmological distances, the short durations and the observed brightnesses of the bursts require that the energy densities in the sources are sufficiently large that photon-photon scattering would hinder escape of radiation unless motions within them are ultrarelativistic. By *ultrarelativistic* we mean that matter moves at speeds that are within a few percent that of light. Peter Mészáros of Pennsylvania State University and Sir Martin Rees of the University of Cambridge have done much to explore the consequences of an ultrarelativistic fireball model of the gamma-ray bursts. In it, the basic flow structure is similar to that of a supernova remnant in its free expansion phase, when the remnant has a two-shock structure (cf. section 8.4). Ordinary supernova ejecta flow at only roughly several percent of the speed of light. Shocks at the higher speeds in the gamma-ray bursts heat gas to temperatures far in excess of those reached in standard young supernova remnants. At such high temperatures the collisions that cause cooling by free-free emission or radiation following electron impact-induced excitation of bound states of ions (cf. section 8.4) are very inefficient; thus, in the gamma-ray bursts, the cooling and emission are due to electrons spiralling in a magnetic field. This mechanism, synchrotron emission, was introduced near the end of section 8.6 to account for emission at radio wavelengths in supernova remnants, but for sufficiently strong magnetic fields and energetic particles it will cause radiation in the gamma-ray range of the electromagnetic spectrum.

The ultrarelativistic gamma-ray burst model was used to predict afterglow at longer wavelengths. Afterglow is seen in the coals of an extinguished fire; their color changes until they have cooled sufficiently that they appear to be black. As the gamma-ray burst source expands, it cools and the magnetic field in it weakens, combining to cause the emitted radiation to take place at progressively longer wavelengths. In the mid-1980s Dutch and Italian scientists began a collaboration to build a satellite that would search for the X-ray afterglows of gamma-ray bursts. The X-ray detectors were constructed to have much better directional resolution than the gamma-ray detector accompanying the experiment to find the burst sources. At the same time the X-ray detectors were designed to be able to observe a large fraction of the sky at any time. Satellites take a long time to build and to launch. Thus, it was only in 1997 when this Dutch-Italian satellite named BeppoSAX provided the first data on the X-ray afterglow of a burst detected in gamma-rays. Subsequently, afterglow at optical wavelengths was also seen. The high directional resolution of the X-ray and optical data permitted the determination of the precise location of the gamma-ray bursts. They were found to be in other galaxies including very distant ones. From the distances to the other galaxies and the measured brightnesses of the gamma-ray bursts, an energy comparable to the *total* energy of a supernova was inferred to be associated with each burst. We have emphasized the word *total* because only about 1 percent of a supernova's energy is usually associated with the motions of the ejecta; most of the energy is lost in neutrino emission. The gamma-ray bursts are possibly some type of supernovae in which energy that in ordinary supernovae goes into neutrinos is deposited into the ejecta which somehow are accelerated to much higher speeds than in normal supernovae. Theoretical understanding of why some supernovae are gamma-ray bursts and others aren't is incomplete, but the evolution of the afterglow of gamma-ray burst sources has now been studied observationally and the sources evolve to be similar to standard supernovae. Parts of the mystery are solved, but many questions remain unanswered.

SELECTED REFERENCES

Bode, M. F., and Evans, A., eds. *Classical Novae.* Chichester: John Wiley & Sons, 1989.

Corradi, R. L. M., and Schwarz, H. E. "Bipolar Nebulae and Binary Stars: The Family of Crabs He 2-104, BI Crucis and My CN 18." *Astronomy and Astrophysics* 268 (1993): 714–725.

Fishman, G. J., and Meegan, C. A. "Gamma-Ray Bursts." *Annual Review of Astronomy and Astrophysics* 33 (1995): 415–458.

Hartmann, L., and Kenyon, S. J. "The FU Orionis Phenomenon." *Annual Review of Astronomy and Astrophysics* 34 (1996): 207–240.

Nussbaumer, H., and Walder, R. "Modification of the Nebula Environment in Symbiotic Systems Due to Colliding Winds." *Astronomy and Astrophysics* 278 (1993): 209–225.

Shore, S. N., Livio, M., and van den Heuvel, E. P. J. *Interacting Binaries.* Berlin: Springer-Verlag, 1994.

van Paradijs, J., Kouvelioten, C., and Wijers, R. A. M. J. "Gamma-Ray Burst Afterglows." *Annual Review of Astronomy and Astrophysics* 38 (2000): 379–425.

EPILOGUE

Chapter 11 of this volume contains brief treatments of a number of topics not given much, if any, consideration in previous ones. However, we still have failed to address all of the interesting astronomical winds and explosions.

We gave even the Sun only superficial attention. Many issues about solar and stellar activity, including the mechanism of magnetic reconnection which drives flares on the Sun and on other stars, would have been worthy of deeper pursuit.

Another type of explosion occurring in the Solar System, unmentioned until now, is that associated with the volcanic activity on Io. Heating due to the tidal interaction with Jupiter maintains a molten and active center in this moon. The volcanoes produce a torus of ionized material, rich in sulphur, in the space between Io and the planet.

Very notably, we have not dealt at all with the elementary-particle physics and nucleosynthesis important at the beginning of the Big Bang (as a consequence of the existence of excellent books on the topic).

The list of related subjects that we have not presented here could continue. We are fortunate that the Universe is so rich in phenomena that inspire our curiosity.

Our omissions are striking, but so, too, is the human ability to understand a huge variety of classes of sources through the use of a few basic models. Many of our descriptions are based on one simple picture represented in figure 4-6. A few other concepts, including those of disks and cooling by the emission of radiation, appeared repeatedly. Much of astrophysics progresses through the realization that unsolved problems bear some relationship to those that have been thoroughly investigated. We hope that the reader will have obtained some insight into how scientific explanations are developed, as well as learned more about a number of different fascinating astronomical sources.

GLOSSARY

We restrict the entries to terms that are either undefined in the text or appear somewhere before they are treated more fully. The reader should also consult the index.

Active galactic nucleus. A source emitting energy at about one billion to one thousand billion times the rate at which the Sun does and having a size less than roughly several times the distance of the Earth to the Sun. Many astronomers expect that sufficiently sensitive observations would reveal all active galactic nuclei to be at the centers of galaxies. However, only a small fraction of galaxies contain active galactic nuclei at the present.

Ambient material. Material that initially surrounds an object that generates a wind or explodes.

Angular resolution. The minimum angle that can exist between the lines of sight to two objects if an observer is to recognize the objects as spatially separated sources.

Astronomical unit. The distance of the Earth from the Sun, which is 150 million kilometers. Its abbreviation is AU.

Aurora. An atmospheric emission phenomenon triggered by the impact of particles ejected by the Sun. The most brilliant aurorae produce optical emission that appears to emanate from a large fraction of the sky. Aurorae are usually observed in polar regions.

B star.   A star with a surface temperature between about 11,000 and 28,000 Kelvin. A star with a mass between about 2.5 and 12 times that of the Sun will have a surface temperature in that range when hydrogen burning has just started in its core. Depending on its mass a B star at the start of the main-sequence phase of evolution will have a luminosity of about 40 to 10,000 times that of the Sun.

Big Bang.   The name associated with a widely accepted model of the Universe's evolution. In that model, the Universe was on average denser at earlier times than it now is and continues to expand. George Gamow used the model to correctly predict the existence of a remnant microwave radiation background. He and his collaborators also used the model to account for the production of helium and deuterium, a "heavy" isotope of hydrogen, in the Universe.

Black body spectrum.   The spectrum of electromagnetic radiation emitted by a totally opaque object. The spectrum depends only on the object's temperature. At higher temperatures the emission is predominantly at shorter wavelengths.

C-N-O chain.   A sequence of nuclear reactions involving carbon, nitrogen, and oxygen that leads to the fusion of hydrogen to form helium and release positrons and neutrinos. There is no net loss or production of carbon, nitrogen, and oxygen due to the sequence.

Continuum emission or radiation.   Radiation emitted over a continuous range of frequencies or wavelengths.

Corona.   A hotter, more tenuous region of material that extends above the surface layer of a star that dominates the emission of optical radiation. In the case of the Sun, much of the coronal material is at temperatures exceeding 1 million K, whereas the surface emitting most of the Sun's light is at less than 6,000 K. During a total solar eclipse the solar corona appears as a bright crown around the Moon.

Cosmological constant.   A term in the equation governing the gravitational field. In his general theory of relativity, Einstein originally assumed the term to be zero. He later introduced it in order to allow him to construct models of the Universe in which neither expansion nor contraction occur. After Hubble discovered that all but the nearest galaxies are receding from us, Einstein dropped the cosmological constant and favored a model of the Universe in which expansion is occurring. The cosmological constant has been the subject of much attention in recent years; many physicists and astronomers suspect that it may be causing the Universe's expansion to be accelerating.

Cosmology.   The study of the large-scale evolution of the Universe.

Dwarf galaxy.   A galaxy much less massive than the Milky Way galaxy in which we live.

Extreme ultraviolet radiation.   Electromagnetic radiation with a wavelength in the range from 91.2 nanometers down to about 10 nanometers. 91.2 nanometers is the maximum wavelength of radiation that can cause the ionization of hydrogen.

Far ultraviolet radiation. Electromagnetic radiation with a wavelength in the range of about 200 nanometers to 91.2 nanometers.

Gamma-ray radiation. Electromagnetic radiation with a wavelength shorter than about 0.01 nanometers.

Gamma-ray bursts. Bursts of short wavelength radiation discovered in the 1970s. Their origins remained debated until near the end of the 1990s. The bursts are now known to arise in other galaxies in explosions that eject material at close to the speed of light.

Infrared radiation. Electromagnetic radiation with wavelengths between about 1 micron (1,000 nanometers) and 300 microns. H_2 infrared radiation with a wavelength near 2 microns is observed from molecular gas in regions where star formation is occurring.

Intergalactic medium. The matter between the galaxies.

Isotope. A form of an element with a specific number of neutrons in each nucleus. For instance, a nucleus of carbon contains six protons. Most but not all nuclei of carbon contain six neutrons each, but some carbon nuclei contain seven neutrons each whereas others contain eight. A carbon nucleus having six neutrons is said to be of one isotopic variety, while one having seven neutrons is said to be of another isotopic variety.

Light year. The distance that light travels in a year; it is about 1×10^{13} kilometers or 60,000 astronomical units.

Luminosity. The power or energy emitted per unit time by a source.

Magnetic storm. A period when electric currents in the Earth's atmosphere cause local variations in the magnetic field. The variations can be sufficient to deflect a compass.

Magnetic reconnection. A mechanism for the rapid conversion of magnetic energy into heat when parcels of material in which the magnetic field directions are very different contact one another.

Main-sequence star. A star that has finished the collapse that forms it and is powered by hydrogen burning in its core.

Maser. A source of radio wave, millimeter wave, or submillimeter wave emission that is much stronger over a narrow wavelength range than it would be in a black body spectrum for the temperature of the material composing the maser.

Maunder minimum. A period in the late 17th century when Sunspots were few in number for much longer than the 11-year period associated with the Sunspot cycle.

Millimeter wave radiation. Electromagnetic radiation with wavelengths between about 1 and 10 millimeters. Submillimeter wave radiation has wavelengths between about 300 microns (0.3 millimeters) and 1 millimeter.

Muon.   A particle with a mass that is about 207 times the mass of the electron, which has a mass that is about 1,840 times smaller than that of the proton. Muons are much less stable particles than protons and electrons but much more stable than many types of subatomic particles. A muon at rest decays into other particles in a couple of microseconds.

Nanometer.   One billionth of a meter or one thousandth of a micron. The abbreviation for nanometer is nm.

Nebula.   Any astronomical source other than a planet that does not appear point-like. Nebulae belong to two general classes, one of which is that of galaxies. Diffuse gaseous objects constitute the other class. Such diffuse gaseous sources include gas ionized by stars that have just formed (cf. color plate 5) or matter that has been ejected by evolved stars (cf. color plate 3).

Neutrinos.   A subatomic particle whose existence was suggested by Wolfgang Pauli. He was explaining the apparent violation of the conservation of momentum in some types of decays of subatomic particles. Subsequent experiments confirmed the neutrino's existence. For many decades the neutrino was believed to be a massless particle and to travel at the speed of light, but recent work shows that it possesses mass and travels slower than but close to the speed of light. A neutrino interacts much more weakly with other matter than most types of particles do, and one could easily escape from the center of a star like the Sun.

Neutron star.   A star composed primarily of neutrons, which are neutral subatomic particles that are only slightly more massive than protons. The central cores of some stars that are supernova progenitors become neutron stars.

Newton's Second Law.   The proposition that the rate per unit time at which the momentum of a body changes is equal to the force on the body.

O star.   A star with a surface temperature above about 28,000 Kelvin. A star with a mass greater than about 12 times that of the Sun will have a surface temperature in that range when hydrogen burning has just started in its core. Depending upon its mass, a young main-sequence O star will be from 10,000 to more than a million times as luminous as the Sun.

Optical radiation.   Electromagnetic radiation with wavelengths between about 350 and 800 nm.

Photon.   A "particle" of light. The concept was introduced by Einstein in his explanation of data showing that electromagnetic radiation must have a wavelength below a maximum in order to cause a material to emit electrons. The maximum depends on the kind of material. Einstein posited that light is composed of discrete units, each of which contains an amount of energy that is inversely proportional to the wavelength of the light. The unit is called a photon. When the energy of a single photon is greater than the energy binding an electron to the material, the emission of an electron can be triggered by the absorption of a photon. Einstein's only Nobel Prize was awarded for his work on this mechanism, which is called the *photoelectric effect*.

π^0. A subatomic particle also called a *neutral pion*. It decays into gamma rays in about only 10^{-16} seconds. Its mass is roughly one seventh that of a proton.

Plasma. A state of matter in which a significant fraction of electrons are not bound to individual atoms. A plasma can flow like a gas.

Positron. A particle having the same mass as an electron but carrying a single positive charge rather than a single negative charge as an electron does.

Proton-proton chain. The set of reactions that fuse protons together to form helium in many of the lowest mass stars. In it protons react to form deuterium, positrons, and neutrinos. A particle of deuterium is composed of a proton and neutron bound together. Deuterium reacts with protons to form an isotope of helium that contains two protons and one neutron in each nucleus; gamma rays are also emitted in the deuterium-burning reaction. The particles of the rare isotope of helium fuse together to form helium nuclei, containing two protons and two neutrons each; protons are formed simultaneously.

Radiation pressure. The force per unit area exerted by the impact of electromagnetic radiation on a surface.

Radio radiation. Electromagnetic radiation with a wavelength of about a centimeter or longer.

Resolution. See *Angular resolution* and *Spectral resolution*.

Resolved source or feature. An entity observed with sufficient angular or spectral resolution for its main characteristics to be distinguished.

Schwarzschild radius. In some sense, the radius of the outer boundary of a nonrotating black hole. Nothing, including light, that passes as close as the Schwarzschild radius to the center of such a black hole can escape.

Shock. The surface at which abrupt changes in a flow's density, velocity, and other properties occur as a consequence of the material in the flow moving at a supersonic speed with respect to an obstacle.

Soft X-ray radiation. Electromagnetic radiation with wavelengths between about one and ten nanometers.

Solar flare. A region on the solar surface that brightens substantially on a timescale that is usually roughly a half hour to several hours. To be classified as a flare the region must have a surface area that is at least a tenth of a thousandth that of the Sun.

Solar mass. About 2×10^{30} kilograms.

Solar radius. About 700,000 kilometers.

Source. A source of radiation. A partial list of astronomical sources includes stars, galaxies, planets, active galactic nuclei, the intergalactic medium, and nebulae around stars.

Spectral resolution. The smallest difference between the frequencies or wavelengths of two beams of radiation that can be measured with a particular apparatus.

Subluminal motion. Motion that is slower than the speed of light.

Sunspot. A dark region on the surface of the Sun. Such a region is dark due to its having a lower temperature than the material surrounding it. The magnetic field in a Sunspot helps isolate it thermally from its surroundings because the motions of charged particles, which carry heat, are restricted by the magnetic field.

Superluminal motion. Motion that is faster than the speed of light.

Supernova. A type of stellar explosion that ejects a good fraction of a star's mass at speeds exceeding several thousand kilometers per second.

Supernova remnant. The object created through the interaction of the ejecta of a supernova and the interstellar matter near it.

Supersonic motion. Movement at a speed exceeding the sound speed.

Type II supernova. A type of supernova that develops when a sufficiently massive star has exhausted the nuclear fuel in its core.

Ultraviolet radiation. Electromagnetic radiation with wavelengths between about 350 nanometers to 10 nanometers. See also *Extreme ultraviolet radiation* and *Far ultraviolet radiation*.

Unresolved source or feature. An entity that is not resolved; see *Resolved source or feature*.

Wolf-Rayet star. A highly evolved massive star with a wind that is considerably more powerful than that of the main-sequence star that was its progenitor. The initial mass of a star that evolves to become a Wolf-Rayet star is at least about 25 times the Sun's mass.

X-ray radiation. Electromagnetic radiation with wavelengths between about 0.01 and 10 nanometers. See *Soft X-ray radiation*.

MATHEMATICAL APPENDIX

Throughout the volume the dependencies of various quantities (e.g., the temperature behind a shock) upon others (e.g., the speed of the shock) are stated in words. This appendix is devoted to the presentation of the relationships between quantities in the form of equations, which may make these relationships clearer to some readers. The relevant section number of material in the main text is given.

2.6

T_S indicates the temperature immediately behind the shock, and V_S is the speed of the shock as measured by someone who is at rest with respect to the medium into which the shock is propagating. Then,

$$T_S = 1.4 \times 10^5 \, \text{K} \left(\frac{V_S}{100 \, \text{km} \, \text{s}^{-1}} \right)^2$$

T_R is the radiation temperature associated with a black body spectrum. λ_{max} is the wavelength at which the energy emission rate per unit wavelength is a maximum. Wien's Law is

$$\lambda_{max} = \frac{2.9}{T_R} \, \text{mm}$$

where T_R is in degrees Kelvin.

F is the total energy radiated per unit area of an object. If its spectrum is like that of a black body of temperature T_R

$$F = \sigma T_R^4$$

where σ is the Stefan-Boltzmann's constant. Max Planck founded the field of quantum theory in 1901 when he developed an explanation of the spectrum of a black body.

3.1

The structure of a stellar interior is usually nearly in a state of hydrostatic equilibrium. In such a state forces balance one another so that no bulk motions occur. If G is the universal gravitational constant, ρ is the mass density, r is the distance to the center of the star, and M is the mass of material interior to the star, the gravitational force in the radial direction on a thin shell of thickness δr is

$$F_G = -\frac{GM}{r^2}\left(\rho 4\pi r^2 \delta r\right)$$

where the minus sign indicates that the force is inward and $\rho\, 4\pi r^2 \delta r$ is just the mass of the material in the shell. The pressure of the gas at radius r is P.

$$P = knT$$

where k, n, and T are the Boltzmann constant, the number density of particles, and the gas temperature, respectively.

At radius $r + \delta r$ the pressure is $P + \delta P$. The net force due to pressure on the shell is

$$F_P = -\delta P\left(4\pi r^2\right)$$

In the star, hydrostatic equilibrium requires that

$$F_G + F_P = 0$$

which leads to the conclusion that

$$\frac{GM}{r^2}\rho = -\frac{\delta P}{\delta r}$$

Those who know calculus will realize that this is the same as

$$\frac{GM}{r^2}\rho = -\frac{dP}{dr}$$

where the term on the right of the equation is the derivative of P with respect to r.

4.1

If U is the thermal energy density, the amount of thermal energy per unit volume, then for an ionized gas or neutral atomic gas

$$P = \tfrac{2}{3} U$$

The thermal speed may be defined in many ways. One commonly used definition is the root-mean-square of the speeds of the gas particles. This definition is used only if the average velocity of the particles is zero—that is, the center of mass of the system of particles is at rest. If a system contains only three particles having speeds of V_1, V_2, and V_3, respectively, the root-mean-square speed is

$$V_{\text{rms}}(3) = \tfrac{1}{3}\sqrt{V_1^2 + V_2^2 + V_3^2}$$

If a fourth particle having a speed of V_4 is added to the system, the root-mean-square speed is

$$V_{\text{rms}}(4) = \tfrac{1}{4}\sqrt{V_1^2 + V_2^2 + V_3^2 + V_4^2}$$

The definition of the root mean square speed can be generalized to any number of particles. Let's take the thermal speed, which we shall designate by V_{TH}, to be equal to the root-mean-square speed. Then

$$U = \tfrac{1}{2} n m V_{\text{Th}}^2$$

where n is the number of particles in a unit volume and m is the mass per particle.

Clearly

$$P = \tfrac{1}{3} n m V_{\text{Th}}^2$$

If u is the bulk speed of a fluid, then the ram pressure, P_{ram}, is given by

$$P_{\text{ram}} = \rho\, u^2$$

where ρ is the mass density.

Bernoullis's Law is

$$P + P_{\text{ram}} = P + \rho u^2 = \text{constant}$$

The sound speed of an ionized gas or a neutral atomic gas that is not emitting or absorbing radiation or gaining or losing energy due to conduction or any number of related processes is

$$a_s = \left(\frac{5}{3}\frac{P}{\rho}\right)^{\!\tfrac{1}{2}}$$

From the equation for a_s and the equations relating P to U and U to V_{Th}, it follows that

$$V_{\text{Th}} = \frac{3}{5}\sqrt{5}\, a_s = 1.3 a_s$$

4.2

The escape speed at a distance r from the center of a spherical object of mass M is

$$V_{\text{esc}} = \left(\frac{2GM}{r}\right)^{\frac{1}{2}}$$

4.3

In a steady constant-velocity, spherically symmetric flow directed radially outward.

$$\rho \propto \frac{1}{r^2}$$

5.2

The magnetic pressure, P_B is given by

$$P_B = \frac{B^2}{2\mu_o}$$

where B is the magnetic field strength and μ_o is a constant called the magnetic susceptibility. Consider a coordinate system with x, y, and z axes. Assume that P_B does not depend on the values of y and z, but does depend on the value of x such that the difference between P_B at $x = \delta x$ and $x = 0$ is δP_B. We assume that δx is small. Then at $x = 0$ the force per unit volume due to magnetic pressure is $-\delta P_B/\delta x$ in the direction of the x axis.

To see how the force per unit volume due to magnetic tension is calculated, consider a case in which at $x = 0$ the magnetic field is of strength B_x and is in the direction of the x axis. At $x = \delta x$, where δx is small, the magnetic field has a small component δB_y in the direction of the y axis but no component in the direction of the z axis. We assume that δB_y does not depend on the values of y or z. Clearly, the magnetic field is bending somewhat from being purely along the direction of the x axis. The force per unit volume due to the tension of the magnetic field is given by $-B_x \delta B_y/\delta x$ and is in the direction along the y axis. The minus sign indicates that the force pushes material toward the x axis.

The Alfvén speed is

$$V_A = \frac{B}{\left(\mu_o \rho\right)^{\frac{1}{2}}}$$

Fast magnetosonic waves propagating perpendicular to the large-scale magnetic field do so at a speed given by

$$V_f = \left(V_A^2 + a_s^2\right)^{\frac{1}{2}}$$

5.3

The angular momentum of a particle of mass m moving in a circle of radius r at a speed v in a counterclockwise direction around the x axis of a coordinate system is

$$L = mvr$$

along the direction of the x axis.

The Jeans mass, M_J for an unmagnetized medium composed of atomic hydrogen is given by

$$M_J = 20 \, \text{solar masses} \left(\frac{T^3}{n}\right)^{\frac{1}{2}} \propto \frac{T^2}{P^{\frac{1}{2}}}$$

where T is given in degrees Kelvin and n is given in terms of the number of hydrogen atoms in a cubic centimeter.

8.4

For a supernova remnant of age t in its adiabatic (i.e., constant energy) phase that is expanding into a uniform surrounding medium with a hydrogen ion number density of n, the outer radius is given approximately by

$$R_{\text{SNR}} = 100 \, n^{-\frac{1}{3}} \left(\frac{t}{10^5 \, \text{years}}\right)^{\frac{2}{5}} \text{light years}$$

n is given in number per cubic centimeter. A typical energy for a type II supernova of 10^{44} Joules ($\equiv 10^{51}$ ergs) has been assumed. R_{SNR} depends on the explosion energy E as

$$R_{\text{SNR}} \propto E^{\frac{1}{5}}$$

INDEX